T0214636

Undergraduate Lecture Notes in Physics

Undergraduate Lecture Notes in Physics (ULNP) publishes authoritative texts covering topics throughout pure and applied physics. Each title in the series is suitable as a basis for undergraduate instruction, typically containing practice problems, worked examples, chapter summaries, and suggestions for further reading.

ULNP titles must provide at least one of the following:

- An exceptionally clear and concise treatment of a standard undergraduate subject.
- A solid undergraduate-level introduction to a graduate, advanced, or non-standard subject.
- A novel perspective or an unusual approach to teaching a subject.

ULNP especially encourages new, original, and idiosyncratic approaches to physics teaching at the undergraduate level.

The purpose of ULNP is to provide intriguing, absorbing books that will continue to be the reader's preferred reference throughout their academic career.

Series editors

Neil Ashby
University of Colorado, Boulder, CO, USA

William Brantley
Department of Physics, Furman University, Greenville, SC, USA

Matthew Deady
Physics Program, Bard College, Annandale-on-Hudson, NY, USA

Michael Fowler
Department of Physics, University of Virginia, Charlottesville, VA, USA

Morten Hjorth-Jensen
Department of Physics, University of Oslo, Oslo, Norway

Michael Inglis
Department of Physical Sciences, SUNY Suffolk County Community College, Selden, NY, USA

More information about this series at http://www.springer.com/series/8917

Albert P. Philipse

Brownian Motion

Elements of Colloid Dynamics

 Springer

Albert P. Philipse
Van 't Hoff Laboratory for Physical
   and Colloid Chemistry, Debye Institute
   for Nano-Materials Science
Utrecht University
Utrecht, The Netherlands

ISSN 2192-4791   ISSN 2192-4805   (electronic)
Undergraduate Lecture Notes in Physics
ISBN 978-3-030-07444-9   ISBN 978-3-319-98053-9   (eBook)
https://doi.org/10.1007/978-3-319-98053-9

This Springer imprint is published by the registered company Springer Nature Switzerland AG
The registered company address is: Gewerbestrasse 11, 6330 Cham, Switzerland

Oedipus, in travelers garb, pondering the riddle of the Sphinx while sitting on a rock. Attic red-figure cup by the Oedipus Painter, c. 470 BC. Rome, Vatican Museum 16541.

Preface

This book nucleated as a set of undergraduate lecture notes. Looking for additional suitable textbook material on Brownian motion, this was often found either too brief and qualitatively or too extensive with more mathematics than desirable for an introductory course. This book aims to be somewhere in between and intends to provide a treatment of Brownian motion on a level appropriate for bachelor students of physics, chemistry, soft matter, and the life sciences.

One very appealing aspect of Brownian motion, as this book also illustrates, is that the subject connects a broad variety of topics, including thermal physics, hydrodynamics, reaction kinetics, fluctuation phenomena, statistical thermodynamics, osmosis, and colloid science. For basic courses on any of these topics, I hope this book will offer useful and motivating study material.

I would like to acknowledge the many insightful discussions with Prof. Agienus Vrij on colloids, Brownian motion, and osmotic pressure. Maria Bellantone (Springer UK) is thanked for the pleasant collaboration—and for encouraging me to finally complete this book. Maria Uit de Bulten-Weerensteijn and Yvette Roman are thanked for their help in the preparation of the book. Samia Ouhadjji and Bonny Kuipers are acknowledged for proofreading. The anonymous referees and Ute Heuser (Springer Physics) have offered helpful comments. Any remaining typos, mistakes or unclarities are, of course, the author's sole responsibility.

Utrecht, The Netherlands Albert P. Philipse
2018

Contents

1   A First Round of Brownian Motion   1
    1.1   A Restless Realm   1
    1.2   Stokes-Einstein Relations   3
    1.3   The Particle Quartet   4
    1.4   Outlook   6

2   A Feverish Sphinx   9
    2.1   Through a Small Grain of Glass   9
    2.2   Molecular Size   12
    2.3   Molecular Reality   13
    2.4   Colloids Are Molecules   15
    2.5   Kinetic Therapy   18
    References   20

3   Kinetic Theory   21
    3.1   The Basis   21
    3.2   Free Volumes and Collisions   23
    3.3   Pressure from Ideal Thermal Particles   27
    3.4   Velocity Distributions and Energy Equipartition   31
    3.5   Soft Matters   42
    References   45

4   A Tale of Ten Time Scales   47
    4.1   Brownian Versus Ballistic Motion   47
    4.2   Mass-Related Time Scales   49
    4.3   The Diffusive Regime   55
    References   60

5   Continuity, Gradients and Fick's Diffusion Laws   61
    5.1   The Continuity Equation   61
    5.2   Constitutive Equations and Fick's Laws   64

5.3   Stationary Diffusion   67
5.4   Diffusion in a Dilute Gas   68
References   70

6   Brownian Displacements   71
6.1   Einstein for Chemists   71
6.2   Translational Diffusion Coefficient from Equilibrium.........   77
6.3   Quadratic Displacements via Einstein's Diffusion Approach ...   80
6.4   Brownian Motion from Newtonian Mechanics   82
6.5   Angular Displacements from a Diffusion Equation   85
6.6   The Rotational Diffusion Coefficient   88
References   91

7   Fluid Flow   93
7.1   Fluid Velocity Fields   94
7.2   The Navier-Stokes Equation   96
7.3   Stokes Flow   100
7.4   On Magnitude   102
References   103

8   Flow Past Spheres and Simple Geometries   105
8.1   Slits and Tubes—and Darcy's Law   105
8.2   Friction Factor of a Rotating Sphere   109
8.3   The Translational Friction Factor   113
8.4   Stick, Slip and the Lotus Sphere   118
References   120

9   Encounters of the Brownian Kind   121
9.1   Diffusion Versus Convection   121
9.2   Brownian Motion Towards a Spherical Absorber   123
9.3   Diffusional Sphere Growth   127
9.4   Birth and Growth of Brownian Clusters   127
References   132

10   Random Walks in External Fields   133
10.1   One-Dimensional Diffusion   134
10.2   Radial Brownian Motion and Colloidal Stability   136
10.3   Brownian Motion in a Shear Flow   138
10.4   Brownian Magnets in a Magnetic Field   139
10.5   Gravity   142
10.6   Exercises   145
References   145

11   Brownian Particles and Van 't Hoff's Law   147
11.1   Thermodynamics of Dilute Solutions   148
11.2   Osmotic Pressure Gauged via the Solvent   150

11.3   Osmotic Pressure from Brownian Motion; Vrij's Statistical
          Approach 151
     References 154

Appendix A: Moments, Fluctuations and Gaussian Integrals 155

Appendix B: Summary Vector Calculus 159

Appendix C: Answers to Selected Exercises 165

Index 175

About the Author

Albert P. Philipse (1956) studied chemistry and philosophy at Utrecht University and achieved his Ph.D. (1987) in colloid science, with a thesis on light scattering from concentrated dispersions of charged silica spheres. At the Energy Research Center (ECN, the Netherlands), he lead projects on colloidal processing of technical ceramics and inorganic foams. In 1991, he joined the Van't Hoff Laboratory for Physical and Colloid Chemistry at Utrecht University where he is Full Professor of Physical Chemistry since 1994. His main research interests are the chemical synthesis, thermodynamics, and physical transport properties of dispersions of charged, magnetic, and non-spherical colloids.

He was Director of Chemical Education in 2006–2009 and is since 2016 Director of the master program Nanomaterials Science of Utrecht University. Albert teaches physical chemistry, thermodynamics, and colloid science on bachelor's, master's, and Ph.D. levels. He also has a teaching assignment in history of chemistry, and delivers guest lectures in primary and secondary schools. In addition to colloid science, Albert's enthusiasms include music, travelling, cooking, and Ancient Greek.

Symbols

Lowercase Greek Letters

α Ratio of maximal energy of a dipole in magnetic field and thermal energy
β Ratio of tangential stress and liquid speed
$\dot{\gamma}$ Shear rate
δ Mass density, distance, unit vector
δ_p Particle mass density
ε Di-electrical constant
ε_j Energy of state j
η Viscosity
θ Angular displacement, polar angle
λ Mean free path
μ Translational particle mobility, dipole moment
μ_r Rotational particle mobility
ν Kinematic viscosity
π Osmotic pressure
ρ Particle number density
σ Standard deviation, shear stress
τ Characteristic time, viscous stress
ϕ Particle volume fraction
φ Reduced electrical potential, azimuthal angle
ψ Stream function

Uppercase Greek Letters

Γ_0 Surface number density of particles
Δ Small difference, small distance
Θ Delay factor
Λ Langevin function
Σ Summation sign
X Initial magnetic susceptibility

Ψ Electrical potential

Ω Number of indistinguishable microstates, angular velocity

Lowercase Roman Letters

a   Capillary radius
c   Concentration
d   Molecular diameter, collision diameter
f   Translational friction factor
f_r Rotational friction factor
g   Gravitational acceleration
g_i Degeneracy
h   Altitude
j   Particle flux per second per unit area
k   Boltzmann constant, rate constant, liquid permeability
l_g Gravitational length
ℓ Momentum relaxation step
m   Particle mass
m^n nth moment of a distribution
n   Number of moles
n_j Number of j-particles
p   Pressure
p_j Probability
r   Displacement by Brownian motion, radial distance
r_{rms} Root-mean-square distance
s   Distance
t   Time
u   Speed, fluid velocity
v   Velocity
v_f Free volume
w   Reversible work
x   One-dimensional displacement, radius
x_j Mole fraction of species j
z   Collision frequency on one target particle, number of surface charges

Uppercase Roman Letters

A   Surface area, aperture area
C   Normalization constant, integration constant
D   Translational diffusion coefficient, rod diameter
D_r Rotational diffusion coefficient
E   Energy
E_{kin} Kinetic energy
F   Force

G	Weight
G(s)	Gaussian distribution of the variable s
H	Magnetic field strength
I	Moment of inertia
J	Angular momentum, particle flux per second
K	Force, equilibrium constant, Kelvin temperature
L	Rod length, distance
L_B	Bjerrum length
M	Molar mass
M_S	Saturation magnetization
N	Number of particles
N_{AV}	Avogadro's number
O	Surface area
P	Momentum
P(x)	Distribution function for variable x
R	Particle radius
R_g	Gas constant
S	Entropy, surface area, slip parameter
T	Absolute temperature, torque
U	Potential energy
U_0	Stationary speed
V	Volume, potential
V_P	Particle volume
Z_c	Effusion rate
Z_B	Number of binary collisions per second

Chapter 1
A First Round of Brownian Motion

1.1 A Restless Realm

A glass marble settles in water and comes to permanent rest at the bottom. However, when the marble is divided into tiny colloidal[1] glass particles, they remain suspended in the water, providing it with a whitish haze (Fig. 1.1). This observation is rather astonishing because the total glass weight has not changed, and since the minuscule glass colloids have the same mass density as the initial marble, one would expect them to sink in water as well. With an optical microscope you can perceive what actually is going on: the glass colloids indeed do not sediment but, instead, move and tumble around in random directions. The serenely settling marble has been pulverized into a pandemonium of small glass bits that senselessly jitter around. It is this turmoil that is designated by the expression 'Brownian motion'.

How long will this Brownian motion endure? Let us imagine that the bottle containing the glass colloids dispersed in water is hermetically sealed and stored in a safe for, say 5000 years—about the age of the Great Sphinx of Giza. Suppose your (very distant) descendent opens the bottle and looks at it through a microscope. She will see exactly the same[2] pandemonium as you did: in their splendid isolation the colloids have continued to dance for ages. Times, by the way, that are insignificant for the air bubbles that have been observed in water inclusions in quartz; here the erratic bubble dance has endured since the Jurassic period from about 200 million years ago, when dinosaurs were roaming our planet.

So Brownian motion truly is a *very* persistent phenomenon. And that confronts us with an enigma: where does the energy comes from that powers this never-ending Brownian movement? The phenomenon is surely at odds with our daily experience:

---

[1] Colloids are particles that have, at least in one direction, a size between a few nano-meters and a few microns.

[2] Assuming that the silica colloids have meanwhile not aggregated by van der Waals attractions which would lead to slower diffusing particle clusters.

© Springer Nature Switzerland AG 2018
A. P. Philipse, *Brownian Motion*, Undergraduate Lecture Notes in Physics,
https://doi.org/10.1007/978-3-319-98053-9_1

Fig. 1.1 Left: glass (silicon dioxide) marbles with a radius of 1 cm, stored in ethanol and arrested in a motionless heap by their weight. Right: colloidal glass spheres with a radius of 100 nm remain suspended in the ethanol due to Brownian motion. The suspension of glass colloids has a hazy appearance due to light scattering by the Brownian particles

in order to keep say, bikes and buggies in motion we have to peddle and push. When we stop doing that, bikes and buggies grind to a halt, because of the resistance between internal parts of a vehicle, and friction between the moving vehicle and its surroundings. This friction comprises the dissipation of mechanical energy as heat, i.e. the irretrievable or *irreversible* distribution of energy over vast numbers of molecules in the surroundings.

That, at least, is the state of affairs in the macroscopic world around us. The microscopic domain of molecules and colloids, in contrast, is an agitated realm where particles do not come to a standstill. The particles *do* experience visious friction, that is to say, they transfer kinetic energy to their environment. However, now the energy donation is *reversible*: in equilibrium diffusing particles gain on average just as much energy as they lose. The never ending self-motion is the manifestation of a system's temperature—which gives rise to the term *thermal motion* and the expression *thermal energy* for the energy associated with this 'heat' motion of particles. For molecules the spontaneous thermal motion is referred to as diffusion[3] whereas for colloids it is usually called Brownian motion; the difference between the two terms is nominal as they both denote thermal motion.

1.2 Stokes-Einstein Relations

The components of solutions include solvent molecules and electrolyte as well as larger solutes such as inorganic colloids, proteins and nano-particles. For all these components diffusion or Brownian motion is the thermal transport mechanism. This makes the *diffusion coefficient* that quantifies rate of diffusive displacements a central parameter in a large set of kinetic processes such as chemical reaction kinetics of molecules and aggregation kinetics of colloids and nano-particles. Diffusion coefficients also figure in processes as diverse as diffusion towards a biological cell, fragrance molecules spreading from your deodorant, and nucleation and growth of crystals and droplets in, respectively, supersaturated solutions and vapors.

Diffusion coefficients have been extensively studied with respect to, among many other things, concentration effects owing to interacting particles, or the influence of a confining medium such as a gel, a porous medium or a capillary. These effects are always compared to a well-understood reference process, namely the free diffusion of a single particle in a liquid, far away from other particles or a wall. For such a free particle the *translational* diffusion coefficient[4] D is given by Einstein's equation[5]

$$D = \frac{kT}{f},\qquad(1.1)$$

where kT is the thermal energy, k is the Boltzmann constant, T is the absolute temperature in Kelvin, and f is the friction coefficient of the particle. This diffusion coefficient determines the rate at which a particle displaces itself by diffusive motions according to an equation also due to Einstein:

$$\langle r^2 \rangle = 6Dt \qquad(1.2)$$

---

[3]From the Latin verb *diffundere* 'to scatter, pour out'.

[4]D without any subscript always denotes a translational diffusion coefficient. D_r, for example, is the rotational coefficient.

[5]Equation (1.1) is sometimes also referred to as the 'Sutherland-Einstein equation' which does justice to the fact that William Sutherland (1859–1911) published (1.1) earlier than Einstein. See: W. Sutherland, "The measurement of large molecular masses", *Australian Association for the Advancement of Science. Report of Meeting*, 10 (Dunedin, 1904), 117–121.

Here $<r^2>$ is the average of the square of the displacement r in time t. The form of Eqs. (1.1) and (1.2) is *independent* of the size and the shape of the diffusing particle. Only the magnitude of the friction factor is determined by particle size and shape.

In this book we will mainly consider the case of a spherical particle with radius R in a Newtonian liquid with viscosity η for which the friction coefficient for translational motion equals the so-called Stokes friction factor:

$$f = 6\pi \eta R \tag{1.3}$$

The combined result

$$D = \frac{kT}{6\pi \eta R}, \tag{1.4}$$

is usually called the Stokes-Einstein (SE) diffusion coefficient for translational sphere diffusion. It allows us, for example, to determine the radius of a colloidal sphere from diffusion measurements on a very dilute dispersion. Concentration effects in dense dispersions or confinement of a sphere in a small geometry lead to (sometimes drastic) deviations from Eq. (1.4). These effects, however, do not concern us here: in Chaps. 6 and 8 we derive the SE equation for a single free sphere and discuss some of its applications to colloidal kinetics in Chap. 9.

In addition to translational diffusive steps, a colloid simultaneously also performs rotations in random directions. The corresponding rotational diffusion coefficient D_r of a single, free sphere has the same form as the translational coefficient in (1.1) be it with a different friction factor:

$$D_r = \frac{kT}{f_r} = \frac{kT}{8\pi \eta R^3} \tag{1.5}$$

Rotational diffusion is of importance to understand, among other things, the response of particles to external fields, a topic addressed in Chap. 10. The alignment of a magnetic or electric dipole moment of particles by an external field is counteracted by rotational Brownian motion which tends to randomize particle orientations, just as translational diffusion randomizes particle positions.

1.3 The Particle Quartet

The particle size below which Brownian motion becomes significant, i.e. observable under an optical microscope, is around 4–5 μm. For comparison, the thickness of a human hair is about 50 μm (Fig. 1.2), and pollen grains of the ornamental plant *Clarkia Pulchella* (Fig. 2.2) have diameters in the range 50–100 μm. Hairs and grains are examples of *a-thermal* or *granular* particles that exhibit no Brownian motion. A fiber in a spider web has a thickness of about 2 μm, which is in the colloidal size

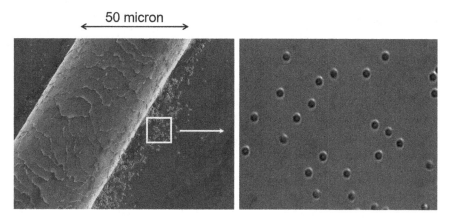

Fig. 1.2 Transmission electron microscope (TEM) image (left) of a human hair sided by colloidal silica spheres of about one micron, observed under an optical microscope (right). A hair is too massive to displace itself by to its thermal energy whereas the silica colloids are small enough to exhibit significant Brownian motion. TEM image courtesy B. Erné and H. Meeldijk; microscopy image courtesy L. Rossi and S. Sacanna

range and thin enough to exhibit thermal fluctuations. Upon decreasing particle size further in the sub-micron range, diffusion becomes progressively more vigorous, in line with the size dependence of the Stokes-Einstein diffusion coefficients in (1.4) and (1.5).

It is instructive to compare, in numerical examples and exercises in this book, four reference particles, spanning seven decades in particle size and twenty decades in molar mass, see Table 1.1. The four will be jointly denoted as the Particle Quartet. The smallest member of the Quartet is the molecular M-sphere, modelling solvent molecules as spheres with a radius of 0.1 nm (Fig. 1.3) which is about the collision radius of a water molecule. The size of the M-sphere represents in order of magnitude also ionic radii in aqueous solutions.[6] Nano-particles are exemplified by the nano N-sphere, which has a radius of 5 nm, a characteristic dimension for metal and metal-oxide nano particles (Fig. 1.3). For the colloidal domain the reference is the colloidal C-sphere with a radius of 100 nm. The largest member of the Quartet is the granular G-sphere with its radius of one millimeter; it stands for granular matter composed of large, visible particles such as sand granules or rice grains.

M-spheres have the molar mass of water (Table 1.1) and the other members of the quartet have a mass density of $\delta = 2.0$ g/ml, which is the density of amorphous silica in glass, and crystalline silica in the form of quartz. The Quartet is, unless stated otherwise, dissolved or suspended in water with viscosity $\eta = 0.89$ cP at a temperature of $T = 298$ K.

Though the references particles from Table 1.1 and Fig. 1.3 have a (nearly) spherical shape, it is important to keep in mind that the expression $D = kT/f$ for the diffusion

---

[6]Y. Marcus, *Ionic Radii in Aqueous Solutions*, Chem. Rev. **88** (1988), 1475–1498.

Table 1.1 Properties of the particle quartet[a]

	R^b (nm)	δ^c_p (g ml^{-1})	M^d (g mol^{-1})
Molecular M-sphere	0.1		18
Nano N-sphere	5	2.0	6.10^5
Colloidal C-sphere	10^2	2.0	5.10^9
Granular G-sphere	10^6	2.0	5.10^{21}

[a]Particles are immersed in water with a viscosity of $\eta = 0.89$ cP ($= 0.89$ mPa.s) and $T = 298$ K
[b]Sphere radius; 0.1 nm is about the collision radius of a water molecule
[c]Mass density
[d]Molar mass; 18 g mol^{-1} is that of water

coefficient, and Einstein's law for the quadratic displacement in (1.2) are, as shown in Chap. 6, independent of particle size and shape. They apply to spheres but equally well to, say, clay platelets, bacteria and proteins.

1.4   Outlook

The outline of this book is as follows. The thermal energy underlying Brownian motion emerges from the kinetic theory of molecules, which is why we start in Chap. 3 with a review of that theory for molecules and colloids in a dilute gas. Molecular speeds from kinetic theory form the starting point for the overview in Chap. 4 of the wide range of time scales underlying Brownian motion for particles immersed in a liquid phase.

The analysis of thermal diffusion rests on the general diffusion equation which, among other things, is explained in Chap. 5, and applied in Chap. 6 to derive the time dependence of quadratic displacements for both translational and rotational Brownian. Chapter 6 also includes Einstein's treatment of Brownian motion 'for chemists', and Langevin's analysis of Brownian movements based on Newtonian mechanics.

The results from Chap. 6 are still independent of the medium in which Brownian motion takes place. Since we are primarily interested in colloids in a liquid phase, Chap. 7 addresses hydrodynamics based on the Stokes equation for viscous flow. The Stokes equation is solved in Chap. 8 for simple geometries and for flow past spheres, the latter to eventually obtain the friction factors for translating and rotating spheres.

Having now available the Stokes-Einstein coefficient for sphere diffusion in a liquid, we apply it in Chap. 9 to processes such as colloidal aggregation, diffusional growth and Brownian motion towards an absorbing sphere. Brownian particles in an external field is the subject of Chap. 10, which addresses the effect of a potential gradient on Brownian motion, including the electrical potential around a spherical target,

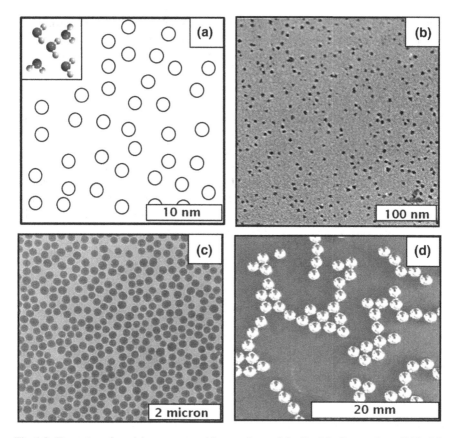

Fig. 1.3 Examples of particles represented by members of the Particle Quartet from Table 1.1.
a Water molecules (insert) are modelled by M-spheres with a radius of 0.1 nm. **b** Cryo-TEM
(Abbreviation of Cryogenic Transmission Electron Microscopy) image of magnetic iron-oxide
particles [M. Klokkenburg et al., J. Am. Soc., **126**, 16706 (2004)] with an average radius of 5 nm,
represented by N-spheres. **c** Cryo-TEM image (A. Philipse & G. Koenderink, *Advances in Colloid
and Interface Science*, **100–102** (2003) 613–639) of amorphous silica spheres with a radius of 90 nm,
slightly below the radius of C-spheres. **d** Stainless steel balls with a radius of one millimeter, equal
to the radius of the granular G-spheres from Table 1.1

the potential energy of Brownian magnets in a magnetic field, and the distribution
of colloids in the gravity field.

The thermal energy that powers Brownian motion also brings about the pressure
that thermal particles exert in the form of gas pressure, or osmotic pressure in a
fluid. The phenomena of osmosis and osmotic pressure are, in their connection to
Brownian motion, the subject of Chap. 11.

An introduction to Brownian motion would be incomplete without attention for
the fascinating story behind the discovery of Brownian motion and its reception
in the 19th and early 20th century. This story, summarized in Chap. 2, is in essence

about resolving the riddle why non-living objects, if small enough, stay in everlasting spontaneously motion. The resolution of this enigma, as it turned out, provided what many scientists considered to be decisive evidence for the existence of molecules.

Exercises

1.1 Calculate for the Particle Quartet in Table 1.1 diffusion coefficients in water at $T = 298$ K.

1.2 How far does each reference particle travel by diffusion in (a) 1 h; (b) one year?

1.3 A glass sphere with a radius of 1 cm is divided[7] into n silica spheres with radius of 100 nm. Compute (a) n and (b) the factor by which the surface area increases.

1.4 Calculate the molar volumes for the Particle Quartet, assuming that each member forms a random sphere packing with a sphere volume fraction of $\varphi = 0.64$.

---

[7] A way to convert a glass marble to nano-silica spheres is to dissolve the marble at alkaline pH to a water glass solution, followed by a slow pH decrease upon which silica nano-particles will nucleate near pH ~ 8. See A. Philipse, *Particulate Colloids* in J. Lyklema (ed.) *Fundamentals of Colloids and Interface Science*, Vol. IV (Elsevier, 2005).

Chapter 2
A Feverish Sphinx

No principal distinction exists, as mentioned earlier, between diffusion of molecules and diffusion of colloids as both represent thermal motion. 'Brownian motion' is the habitual term for colloids; the naming is appropriate as it was Robert Brown who was the first to publish on systematic observations of colloids in motion. Below we will outline Brown's findings and summarize their history of reception, with a crucial role for the kinetic theory of matter that was developed in the second half of the 19th century.

2.1 Through a Small Grain of Glass

The Scottish botanist Robert Brown (1773–1858) was already in his own time well-known as an expert observer with the single-lens microscope. With this modest instrument, essentially a miniature magnifying glass (Fig. 2.1), Brown not only identified the cell nucleus but also studied the fertilization process in plants, for which purpose he investigated the white pollen of the ornamental plant *Clarkia Pulchella* (Fig. 2.2). In June 1827 he observed under his microscope the zigzag motion of tiny objects[1] in water which had escaped from the pollen grains. Such motions, of course, could be expected for small organisms which, in analogy with bacteria or spermatozoa, move by themselves in water without any external assistance. Brown decided to investigate the significance of these zigzagging organisms for the love life of *Clarkia Pulchella* in more detail. Soon, however, he started to doubt whether the tiny, feverishly moving particles were indeed living organisms, even though their motions did not seem to stop. For Brown also scrutinized finely powdered *inorganic* substances (silica, clay, grains of sand) under his single-lens microscope and found that also inorganic particles, if sufficiently small, exhibit erratic motions when

---

[1] These particles are now known to be membrane-bound structures called organelles, including amyloplast which store starch. They have nothing to do with plant fertilization.

© Springer Nature Switzerland AG 2018
A. P. Philipse, *Brownian Motion*, Undergraduate Lecture Notes in Physics,
https://doi.org/10.1007/978-3-319-98053-9_2

Fig. 2.1 Brown used a microscope of this type for his study of Brownian motion. This microscope has only one lens in the form of a small glass grain. Courtesy Dr. J. Deiman, Utrecht University Museum, photograph J. den Boesterd. For the non-Dutch reader: you see the end of a roll of peppermint candies which are discs with a radius of one centimeter

dispersed in water. A mineral also pulverized and scrutinized by Brown was granite, in one case from a remarkable, certainly *very* lifeless source. In his own words:

> Rock of all ages, including those in which organic remains have never been found, yielded the molecules[2] in abundance. Their existence was ascertained in each of the constituent minerals of granite, a fragment of the Sphinx being one of the specimens examined.

Brown's startling conclusion was that inanimate matter *spontaneously* moves in a liquid, provided matter particles are small enough. This conclusion was controversial for many years—and, remarkably, in some corners the possibility that inanimate matter has self-motion is still ruled out[3]. Especially the spontaneity of the particle motion was contested in view of factors such as mechanical vibrations, solvent evaporation, and liquid convections, which could cause the observed motion of suspended particles. Such objections are not unreasonable; dust particles are seen to whirl around in sunlight due to air convection, and even minute temperature gradients set up liquid flows in dispersions.

The author has verified by himself, employing the microscope in Fig. 2.1 using an aqueous dispersion of mono-disperse latex spheres (diameter about one micron), that Brown indeed must have been able to observe colloids in motion. The irregular, diffusive movements of individual latex particles can be distinguished under the microscope of Fig. 2.1, be it with some difficulty, from convective motions due to liquid flow, in which particles jointly move in the same direction.

Such a present-day experiment, of course, not only employs our modern, mono-disperse latex particles in a clean solution, but surely is also guided by what we expect

---

[2]Brown employs the term 'molecules' to denote small self-moving particles.

[3]*"Spontaneous movement is a characteristic sign of life [...] Such a movement is never exhibited by non-living objects"*. A. C. Dutta, *A Class-Book of Botany* (17th "new revised edition", Oxford, 2000)

Fig. 2.2 The ornamental
flower *Clarkia Pulchella* (H.
W. Richett, *Wild Flowers of
the United States*, vol. 6 (Mc
Graw Hill, New York 1967)

to see. An unprejudiced 19th century observer trying to repeat Brown's experiments
must, apart from external disturbances, have been easily confused by the unclear
image of moving and stagnant objects (dust, bacteria, cells, colloids of various size
and shape etc.) observed in a drop of sap or water under a microscope. It is quite
difficult to interpret—or sometimes even to put into words—observations without
guidance by any model or theory.

Father Delsaulx on mechanical theory. That guidance took a long time: it lasted
almost fifty years before Brown's observations were linked to kinetic[4] theory—the
theory that will be reviewed in Chap. 3. The concept of molecules in thermal motion
was central to the kinetic theory that was developed in the second half of the 19th
century. However, making a connection with microscopically observable Brownian

---

[4]From the Greek *kinetikos* derived from the verb *kinein* 'to move'.

motion in a liquid was anything but obvious. It was in any case not obvious to James Clark Maxwell (1831–1879), one of the founders of kinetic theory, who remarked[5]:

> Diffusion in liquids and gases is the strongest evidence that they contain molecules in a state of continuous agitation (these motions) cannot be directly observed.

Christian Wiener (1826–1896) studied Brownian motion by optical microscopy on what we now call a colloidal silica sol, which he prepared by precipitation of an aqueous silicic acid solution. Wiener noted that the motion of the silica particles was too erratic to be caused by liquid convections or mechanical undulations and made in his 1863 publication an attempt to relate Brownian motion to inherent fluctuations of the suspending fluid.

More than a decade later the Belgian Jesuit Rev. Joseph Delsaulx (1877) wrote that

> The motions discovered by Robert Brown in minute particles, and for that reason called Brownian motions, have since been observed by all naturalists. In fact, there is not one amongst them but must have been struck by the strangeness, the persistence, and the frequent apparition of these molecular motions in the field of the microscope; not one, I fancy, who has not tried to raise up, were it only by a corner, the veil which nature has cast upon the secret of their origin. Hitherto, it must be confessed, all their efforts have been fruitless: the Sphinx[6] has kept its enigma.

Father Delsaulx then argues towards the following resolution of the conundrum:

> After having explained, in conformity with the principles of thermo-dynamics, the movement of the Brownian gaseous bubbles, and the little masses of vapor in quartz, I shall endeavor in the same way to account for the movements observed in viscous globules, and solid granulations in liquids. According to me, all these movements result from the interior dynamic state that the mechanical theory of heat attributes to liquids.

Delsaulx also notes that Brownian motion is a remarkable confirmation of this mechanical theory. This confirmation remained qualitative, if not speculative, until statistical thermodynamics came on stage, and until it was clearly apprehended that large particles (colloids) obey the same statistical laws as molecules. This realization was a turning point in a long-standing controversy on the status of atoms and molecules.

2.2 Molecular Size

Everything is composed of molecules that, in turn, are made from the atoms of the Periodic Table. We are familiar with molecules—and easily envisage water molecules being pushed out of the way when stirring a cup of tea, and imagine that we are being

---

[5]J. C. Maxwell, *Theory of Heat* (1888). Unabridged republication by Dover (2001).

[6]The Sphinx (see cover frontispiece) was a horrendous female monster with a women's head, the body of a lion and wings of a bird. She ravaged the city of Thebes, devouring anyone who failed to correctly resolve her riddle.

bombarded by air molecules when cycling against the wind. To confirm, in case of any doubt, that molecules are really there, one can look at their images made with a scanning force microscope. Yet, verifying the reality of molecules without such a powerful microscope, depending only on perception of the macroscopic world, is quite a different matter. Can you mention visual support for air being composed of thermally agitated molecules—or report any evidence for molecular dimensions?

Air molecules. A first credible estimate of the size of 'air molecules' was made by Joseph Loschmidt (1821–1895). It did not come from any visual observation, but was inferred from (Exercise 2.7) Maxwell's kinetic gas theory. Briefly, Loschmidt combined the mean-free-path of gas molecules (Chap. 3, Eq. 3.13) and Maxwell's gas viscosity (Chap. 5, Eq. 5.31) to obtain a result with two unknowns: a molecular diameter d and a gas volume fraction φ. He then estimated the latter from the volume the gas would occupy when condensed to a liquid, which eventually lead to a diameter of order $d \sim 1$ nm for an 'air molecule'.

Franklin calms the waves. There actually *was* visible evidence available on molecular size, almost one century before Loschmidt's 1865 publication. Benjamin Franklin (1706–1785) reported[7] in 1773 that a tea spoon of oil poured into a pond could spread to an extensive oil film that calms down the water. Assuming the film is a mono-layer of 'oil molecules' we can estimate their size from Franklins data. Franklin reports an oil area of about $A = 2000$ m^3 and taking $V = 2$ mL as the typical volume for a Victorian tea spoon we arrive at

$$d = \frac{V}{A} \approx \frac{2 \times 10^{-6} \text{ m}^3}{2 \times 10^3 \text{ m}^2} \approx 1 \text{ nm} \tag{2.1}$$

That is in order of magnitude the length of fatty acid chains of soap molecules in a film at the water-air interface. Curiously enough, Franklin himself did not made the estimate (2.1). Only much later Lord Rayleigh performed the calculation[8] for a film of olive oil on water to find $d \approx 1.6$ nm.

2.3 Molecular Reality

The findings of Loschmidt and Rayleigh apparently did not form the proof that could convince sceptics such as the physicist Ernst Mach (1838–1916), who admitted that molecules were a very useful hypothesis, but anyhow a hypothesis. Wilhelm Ostwald (1853–1932) rejected the reality of molecules, being convinced that all science should be based on phenomenological thermodynamics. There is consistency in his view point: the validity of the First Law (total energy is conserved in any process) and the Second Law (total entropy cannot decrease) does not rely on any particular molecular

---

[7]Phil. Trans. Roy. Soc. **64**, (1774), 445. See also C. H. Giles, *Franklin's tea spoon of oil*, Soc. Chemistry & Industry, Nov. 8, 1969.

[8]Proc. Roy. Soc. (London) **47**, (1890), 364.

microscopic model. So strict adherence to phenomenological thermodynamics is compatible with denying the existence of molecules.

Chemical atoms. 19th Century chemists were drawing schematic diagrams, chemical formulae and stoichiometric equations, since Dalton (1766–1844) introduced his atomic theory. For a critical 19th century student, however, the physical evidence that such chemical symbols might represent 'real' particles was not very convincing. The student could point in the first place to the confusion about the nature of such particles. Were they indivisible atoms[9] in the strict sense of the word? Or were they agglomerates of such atoms? And was there only one type of atom, for example hydrogen as postulated by Prout (1785–1850), or could there be a whole family of chemical atoms - namely one for each element—as advocated by Dalton? Our 19th century student could also point out that, if molecules existed, nobody knew how to count them.

The Boltzmann distribution. In contrast to the phenomenological thermodynamics mentioned above, statistical thermodynamics imposes commitment to a molecular model: it builds up the Second Law on classical mechanics and probability theory, applied to a collection of *discrete* particles in disordered motion. Named after Ludwig Boltzmann (1844–1906), a founding father of statistical mechanics, is the Boltzmann constant[10] k that relates entropy S to probability Ω:

$$S = k \ln \Omega, \tag{2.2}$$

In this equation, carved on Boltzmann's tombstone in Vienna (Fig. 2.3), Ω is the number of indistinguishable microscopic states that correspond to a certain macroscopic state with fixed total energy. We note in passing that the Boltzmann constant is the ratio of the molar gas constant R_g to Avogadro's number:

$$k = \frac{R_g}{N_{AV}} = 1.38 \times 10^{-23} \ JK^{-1}, \tag{2.3}$$

and has the dimension of entropy. Boltzmann's entropy formula has an important consequence for the distribution of an assembly of N particles in an isolated system. According to the Second law, the entropy in an isolated system must increase until equilibrium, that is the state with maximal entropy, is reached. For N particles the maximum of the entropy function $S = k \ln \Omega$ is reached when the particles adopt the Boltzmann distribution:

$$\frac{N_i}{N} = \frac{g_i \exp[-\varepsilon_i/kT]}{\sum_i g_i \exp[-\varepsilon_i/kT]} \tag{2.4}$$

---

[9]From the Greek *a-tomos* 'un-cuttable'.

[10]This constant was actually not employed by Boltzmann, see also Fig. 2.3.

Fig. 2.3 The tombstone of Ludwig Boltzmann (1844–1906) on the *Zentral Friedhof* in Vienna is decorated with Eq. (2.2)— which Boltzmann curiously enough never wrote down. The Boltzmann constant k was introduced in 1901 by Max Planck who also formulated $S = k \log W$

Here N_i is the population at the energy level ε_i, with a degeneracy g_i; the sum in the denominator over energy levels is the partition function of a particle. Such results of statistical thermodynamics are clearly only meaningful when there are particles 'out there', particles that are in thermal motion such that they can evolve to and remain in the equilibrium distribution of Eq. (2.4).

2.4 Colloids Are Molecules

The experimental verification of Eq. (2.4), however, presents a problem. One cannot directly count molecules in such a distribution by, for example, microscopic observations. Yet it was realized by Albert Einstein (1879–1955) and Jean Perrin (1870–1942) that the Boltzmann distribution not only applies to atoms or molecules: it equally holds for the much larger particles in a colloidal suspension (see Fig. 2.4). The reason is that the principle of "equipartition of energy" does not distinguish the thermal motion of a solvent molecule from that of a suspended colloid. The kinetic energy E_{kin} of a particle with mass m translating with a speed u is

Fig. 2.4 Perrin's
microscopic image of the
sedimentation-diffusion
equilibrium of resin spheres
(diameter one micron) in
water. *Source* F.
Randriamasy, Revue du
Palais de la Découverte 20,
no. 197, 18–27 (1992)

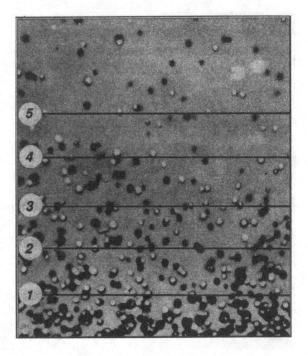

$$E_{kin} = \frac{1}{2}m\,u^2 \tag{2.5}$$

The equipartition principle guarantees that in thermal equilibrium all components of a solution (solvent molecules as well as colloids, polymers or any other particles) have the same *average* translational kinetic energy, which is fixed by the absolute temperature T:

$$\langle E_{kin} \rangle = \frac{3}{2}kT, \tag{2.6}$$

a result that is derived and further discussed in the review of kinetic theory in Chap. 3. Thus the root-mean-square speed of a particle is:

$$\sqrt{\langle u^2 \rangle} = \sqrt{\frac{3kT}{m}}, \tag{2.7}$$

showing that at a given temperature colloids with their large masses move slower than molecules.

Barometric profiles. This large mass also 'compresses' the Boltzmann distribution in the earth gravity field. At a height h above the surface of the Earth at $h=0$, the potential energy of a particle with buoyant mass Δm is:

$$U = \Delta mgh, \tag{2.8}$$

where g is the acceleration of gravity. The Boltzmann distribution (2.4) for colloids in the gravity field leads to an equilibrium profile of the form (Exercise 2):

$$\rho(h) = \rho(h = 0) \exp \left[\frac{-\Delta mgh}{kT} \right] \tag{2.9}$$

Here ρ is the colloid number density; $h=0$ denotes the reference plane where $U=0$. This exponential or 'barometric' distribution for non-interacting particles[11], has a thickness characterized by the 'gravitational length':

$$l_g = \frac{kT}{\Delta mg}, \tag{2.10}$$

which is the height at which the number density has dropped to $\rho(h)=\rho(h=0)/e$. For oxygen molecules l_g is several kilometers, whereas colloidal spheres may adopt equilibrium profiles of only several cm or less (Exercise 2.2). Since such spheres can be observed with an optical microscope, Perrin (Fig. 2.4) was able to directly count the number densities predicted by Eq. (2.9). He determined the mass of his colloidal spheres from measurements of their Stokes sedimentation velocity (see Chap. 6). When $\rho(h)$ in Eq. (2.9) is measured as a function of height h at a given temperature, the Boltzmann constant k remains the only unknown. Perrin thus determined experimentally k, and found in this way a reasonable value of $N_{AV} \approx 6 \times 10^{23}$/mol for Avogadro's number.

Monodisperse colloids. Perrin initiated the use of 'well-defined colloids' to study molecular statistics on a spatial scale which is accessible to an optical microscope. In colloid science this 'upscaling' is still an important strategy, and one is still wrestling with the problem that also Perrin had to face: colloidal particles always have a certain distribution in shape and mass (they are 'polydisperse') whereas atoms are monodisperse—if one disregards isotopes. The distribution in Eq. (2.9), however, presupposes particles with identical mass m. Perrin used fairly monodisperse latex spheres, obtained from laborious fractionation procedures on natural latex ('gamboge') solution: by repeated sedimentation a few hundred milligrams of spheres were obtained from one kilo of rubber. Nowadays well-defined colloids can be prepared by precipitation or polymerization of insoluble substances in a solution.

The equivalence between colloids and molecules also lead Perrin to another microscopic determination of Avogadro's number, based on Einstein's equations for diffusive displacements.

The wonder year. Albert Einstein (1879–1955) investigated Brownian motion primarily to develop arguments to support the existence of molecules, and to validate the applicability of statistical thermodynamics. In his own words[12]:

---

[11]For (2.9) to be valid, particle should not only be ideal, but also uncharged, see Sect. 10.5.

[12]A. Einstein, *Autobiographical Notes*, in Paul A. Schilpp (Ed.) *Albert Einstein: Philosopher-Scientist* Vol. 1, p. 47; The Library of Living Philosophers (Open Court Company, 1969).

My major aim in this was to find facts which would guarantee as much as possible the existence of atoms of definite size. In the midst of it I discovered that, according to atomistic theory, there would have to be a movement of suspended microscopic particles open to observation, without knowing that observations concerning the Brownian motion were already long familiar.

In the *annus mirabilis* 1905 (in which he also first published on special relativity and the photo-electric effect) Einstein reported equations for the diffusion of a particle in a liquid, already mentioned in Chap. 1, namely expressions for the diffusion coefficient and the quadratic displacement. Perrin verified Einstein's predictions by measuring the displacements of his colloidal latex spheres under a microscope and, found, via the Stokes-Einstein relation $D = 6\pi \eta R/kT$ again a reasonable value for Avogadro's number. Perrin's experiments made quite an impact; even a sceptic such as Wilhelm Ostwald accepted eventually the reality of molecules on the basis of Perrin's experiments.

Inherent motion. Textbooks sometimes mention Brownian motion as being caused by 'uncompensated' collisions of solvent molecules, which kick around an otherwise inert colloidal particle. The point is, however, that all free, small entities are in inherent thermal motion regardless of their surroundings with which they equilibrate. The colloid's environment may be a liquid, a gas or a bath of electro-magnetic radiation - and in any environment colloids move randomly about an equilibrium position due to their kinetic energy. Only the *distance* they move is determined by the energy dissipation to its surroundings; a viscous damping in case of a liquid. The Stokes-Einstein diffusion coefficient $D = kT/6\pi \eta R$ summarizes this state of affairs: diffusion is driven by the thermal energy kT with no reference to the surroundings of colloids, and damped by the Stokes friction factor which specifies that the colloids in question are suspended in a continuous fluid with viscosity η.

Quanta. Brownian motion not only offered credibility for the reality of molecules. The above mentioned case of Brownian motion in a radiation bath provided—in the hands of Albert Einstein—support for the reality of light quanta as well. Einstein discussed Brownian motion by a reflecting mirror suspended in a space filled with electromagnetic radiation. He came to the insight that the haphazardly fluctuating mirror does not exchange energy with continuous light waves but, instead, with radiation that consists of an ideal gas of particles in the form of mutually independent light quanta.[13]

2.5 Kinetic Therapy

The awareness that the world is composed of particles or atoms in spontaneous everlasting motion has a long history, and one ancestor of kinetic theory deserves to have the floor for a moment. Not only because he foreshadows atomism in a

---

[13] How he came to this insight is eloquently related by Einstein himself in A. Einstein in: Paul A. Schilpp, *op. cit.* p. 51.

remarkable way but also because he does so via beautiful poetry with many captive images of atoms and their motions. The Roman poet Titus Lucretius Carus (*c.*100-*c.*55 B.C.) writes in his *On the Nature of the Universe* on the process of atoms clustering and falling apart:

> This process, I might point out, is illustrated by an image of it that is continually taking place before our very eyes. Observe what happens when sunbeams are admitted into a building and shed light on its shadowy places. You will see a multitude of tiny particles mingling in multitude of ways in the empty space within the light of the beam, as though contending in everlasting conflict, rushing into battle rank upon rank with never a moment's pause in a rapid sequence of unions and disunions. From this you may picture what it is for atoms to be perpetually tossed about in the illimitable void. To some extent a small thing may afford an illustration and imperfect image of great things. Besides, there is a further reason why you should give your mind to these particles that are seen dancing in a sunbeam: their dancing is an actual indication of underlying movements of matter that are hidden form our sight. There you will see many particles under the impact of invisible blows changing their course and driven back upon their tracks, this way and that, in all directions. You must understand that they all derive this restlessness from the atoms. It originates with the atoms, which move of themselves. Then those small compound bodies that are least removed from the impetus of the atoms are set in motion by the impact of their invisible blows and in turn cannon against slightly larger bodies. So the movement mounts up from the atoms and gradually emerges to the level of the senses, so that those bodies are in motion that we see in sunbeams, moved by blows that remain invisible.

Lucretius's atomistic elegy is not a kinetic theory in our sense of the word, with the aim to explain the macroscopic world in terms of underlying microscopic entities. Its principle goal is to liberate humanity of anxiety for death and fear for the supernatural. The world and everything in it is governed by mechanical laws of the atoms and not by the Gods. By believing this, men can live a peaceful and happy life. What more could any theory wish to achieve?

Exercises

2.1 It is often claimed that Brown observed the Brownian motion of pollen grains. Pollen grains of *Clarkia Pulchella* have diameters in the range 50–100 μm. Suppose you would observe the grains in water under a microscope for one hour: calculate the diffusive displacement of a 50-μm grain in that hour. Conclusion?

2.2 (a) Derive the barometric height distribution in Eq. (2.9). Start with formulating the force balance on the particles in the equilibrium profile. (b) How large is l_g for oxygen molecules, and for colloidal spheres with a radius $R = 100$ nm and mass density of 2 g cm^{-3}?

2.3 Explain why Brownian motion does not stop due viscous friction between colloid and surrounding solvent.

2.4 One could argue that a colloid receives heat from its environment and converts it to work of motion, and *vice versa*. However, according to the Second Law it is impossible that heat is fully convert to work in a closed cycle. Does Brownian motion contradicts the Second Law?

2.5 Discuss in how far the quote from Lucretius in Sect. 2.4 can be seen as a description of Brownian motion.

2.6   **Avogadro's number**. Describe the two methods employed by Perrin- and check his corresponding calculations- to find Avogadro's number N_{AV}. Describe at least four other methods to determine N_{AV} and explain for each case the underlying principle.

2.7   **Loschmidt on molecular size**. The estimate of molecular diameter d in 1865 by Loschmidt is based on the proportionality $d \sim \phi\lambda$, where λ is the mean free path length of molecules in a gas in which molecules occupy a volume fraction ϕ.

   (a)   Verify that this proportionality is correct (hint: consult Chap. 3)
   (b)   Find out how Loschmidt was able to deduce or estimate ϕ.
   (c)   Estimate d for nitrogen gas using the proportionality $d \sim \phi\lambda$
   (d)   Calculate Avogadro's number on the basis of your estimate of d.

References

Brown reported his observations in: R. Brown, *A Brief Account of Microscopical Observations Made in the Months of June, July, and August 1827, on the Particles contained in the Pollen of Plants; and on the General Existence of active Molecules in Organic and Inorganic Bodies*, Edinburgh New Philosophical Journal 5 (1828) 358–371; The Philosophical Magazine and Annals of Philosophy Series 2, 4 (1829) 161–173. R. Brown, *Additional Remarks on Active Molecules*, Edinburgh New Philosophical Journal 8 (1829) 314–319; The Philosophical Magazine and Annals of Philosophy Series 2, 6 (1829) 161–166.

For a reconstruction and thorough analysis of Brown's observations, see P. Bearle, B. Collett, K. Bart, D. Bilderback, D. Newman and S. Samuels, *What Brown saw and you can too*, Am. J. Phys. **78** (12) (2010) 1278–1289.

Perrin's book remains an example of engaging and lucid science writing: J. Perrin, *Atoms* (London: Constable & Company, 1916) Transl. D.L. Hamminck.

For an annotated translation of J. Loschmidt, *Zur Grösse der Luftmoleküle*, Sitzungsberichte der Kaiserlichen Akademie der Wissenschaften in Wien 52 (1865) 395–407, see: W.W. Porterfield and W. Kruse, *Loschmidt and the Discovery of the Small*, J. Chem. Education 72 (1995) 870–875.

The contributions of Wiener and Delsaulx: Chr. Wiener, *Erklärung des atomistischen Wesens des tropfbar-flüssigen Körperzuständes, und Bestätigung desselben durch die sogenannten Molecularbewegungen*, Annalen der Physik und Chemie 118 (1863) 97–94. J. Delsaulx, *Thermo-dynamic Origin of the Brownian Motion*, The Monthly Microscopical Journal 18 (1877) 1–7.

For the blood clotting story of Oedipus and the riddle of the Sphinx, see f.e. Jenny March, *Dictionary of Classical Mythology* (London: Cassell, 1999).

The quote from Lucretius is from: *On the Nature of the Universe*, translated by R. E. Latham (Penguin Books, 1986), p. 63–64. For a verse translation see: *On the Nature of the Universe*, A New Verse Translation by Ronald Melville (Clarendon Press, Oxford, 1997).

Chapter 3
Kinetic Theory

Random motion of particles, so well put into words by Lucretius in Sect. 2.5, was quantified by the kinetic theory of matter in the second half of the nineteenth century. As we have seen in Chap. 2, this theory eventually lead to the demasking of Brownian movements as an instance of thermal motion, be it a remarkable instance, that can be observed with microscopy on colloids dispersed in a liquid. This Chapter introduces kinetic theory for thermal particles, for the computation of, among other things, their kinetic energies and the pressure they exert. The magnitude of thermal energy in comparison to chemical bond energies will also lead us into an aside on *soft matter*, the materials that are particularly susceptible to thermal energy at room temperature.

Colloidal smoke. We will address particles in a gas phase, though all results from kinetic theory also apply to the solutes and solvent molecules in a solution. In addition, from the kinetic view point there is no principle difference between a dilute gas of visible colloids and a vapor of invisible molecules. The distinction is one of degree: for a given temperature average motional energies of molecules and colloids are the same, so the heavy colloids move at much lower speeds than molecules.

Gases of colloidal particles, incidentally, are quite common systems and occur in the form of smoke—which is blueish or white owing to the light scattering from the sub-micron particles in the 'aerosol'. Other examples of colloidal vapors are air polluting smog, hair sprays and deodorant, and the suspended colloidal water droplets better known as 'mist'.

3.1 The Basis

Kinetic theory is the theory that explains properties of matter in terms of the motions of its constituent particles; in what follows 'particles' is the joint term for molecules as well as colloids or nano particles. As even dilute gases contain immense numbers of particles that are in ceaseless chaotic motions, their modelling seems like a daunting, if not impossible task. However, as it turns out, the kinetic theory is able to make quite

© Springer Nature Switzerland AG 2018

A. P. Philipse, *Brownian Motion*, Undergraduate Lecture Notes in Physics,
https://doi.org/10.1007/978-3-319-98053-9_3

accurate predictions on the basis of three minimal assumptions (outlined below) on particle motion, particle interactions and particle size.

1. **Random motions**. Particles are in a state of perpetual random motion, moving erratically around, with motional energies that depend only on temperature. This chaotic pandemonium will persist forever because particles cannot cease moving since a temperature of zero Kelvin cannot be achieved. Thermal particle motion has on average no preferred orientations. Consequently, gases and solutions are *homogeneous* on macroscopic length scales, i.e. on scales very much larger than a particle radius. Statistical properties are the same in all directions. For example, a calculation of the average particle speed in any particular direction, as performed in Sect. 3.3, is a calculation for *all* directions.

2. **Collisions only**. Particles are assumed to be mutually independent, which is to say that they behave *ideal*. Mutual independence of particles entails that their potential energy (their energy due to their position with respect to each other) is assumed to be zero. The total energy of particles is therefore the sum of their kinetic energies. The only interactions between particles are elastic collisions, i.e. collisions that do not dissipate any kinetic energy of the particles involved.[1] A cannon ball falling onto a hard floor will bounce forever if the collision between ball and floor is purely elastic; an extreme counter-example is the non-elastic impact of a lump of clay with the floor, where all the lump's kinetic energy is dissipated during one collision.

3. **Low volume fractions**. Particles are not point-like but have a finite volume otherwise they could not collide. It is by collisions that molecules exchange motional energy such that they evolve to the equilibrium velocity distribution to be discussed in Sect. 3.4. However, molecule volumes are assumed to be small enough for the total particle volume to be negligible in comparison to the vessel volume. The volume fraction φ of particles with volume V_p is defined as:

$$\phi = \frac{N}{V} V_p = \rho V_p \tag{3.1}$$

Here $\rho = N/V$ is the number density of N particles in a total system volume V. A small volume fraction $\varphi \ll 1$ implies for particles in a gas that they travel in a straight line (the free path) many times their own diameter between two collisions. In contrast, Brownian particles or solute molecules in solution are surrounded by a dense population of solvent molecules so here the free path is very much smaller, namely of the order of the size of a solvent molecule.

Equilibrium and particle speeds. Ideal gases and colloidal solutions that are in thermodynamic equilibrium harbor uniformly dispersed particles and colloids in, respectively, a vessel and a solution. According to the postulates mentioned above, particles keep on ceaselessly and randomly colliding with each other. Since collisions alter relative particle speeds in a random manner, particles speeds maintain in

---

[1] Also collisions between a particle and container walls are assumed to be purely elastic.

equilibrium a time-independent distribution, also known as the *Maxwell-Boltzmann distribution*, to which we return in Sect. 3.3. So the supposition that in equilibrium all particle speeds are the same would be mistaken: particles that all have the same speed actually represent a *non*-equilibrium state in which entropy would increase because speeds would spontaneously redistribute themselves until the equilibrium distribution has been reached. The implication is that analysis of particle kinetics must focus on *average* values of speeds and kinetic energies for large numbers of particles.

In what follows we should clearly distinguish particle speeds from velocities; the velocity of a particle is a vector with three Cartesian components in the x, y and z direction:

$$\vec{v} = (v_x, v_y, v_z) \tag{3.2}$$

The speed of a particle is the absolute magnitude, or modulus of the velocity vector, which will be denoted by the symbol u:

$$u = \left| \vec{v} \right| = \sqrt{v_x^2 + v_y^2 + v_z^2} \tag{3.3}$$

Since particle speeds are distributed we will be dealing with averages such as:

$$< u^2 > = < v_x^2 > + < v_y^2 > + < v_z^2 > \tag{3.4}$$

Here the brackets denote a number-average over all particles. For example, the average over the square of the velocity component v_x is defined by:

$$< v_x^2 > = \frac{\sum_{j=1} n_j v_{x,j}^2}{\sum_{j=1} n_j} = \frac{1}{N} \sum_{j=1} n_j v_{x,j}^2 \tag{3.5}$$

Here n_j is the number of particles with a velocity component $v_{x,j}$; note that the sum in the denominator represent the total number, N, of particles. We will return to averages as in (3.5) and how they can be evaluated using distribution functions in Sect. 3.3.

3.2   Free Volumes and Collisions

The first quantity we will calculate employing kinetic theory is the average distance a particle traverses between two collisions with other particles, an average referred to as the particle's *mean free path*. It can be computed from two different expressions for the same statistical property of particles and that is their mean free volume.

Mean free volume. The free volume v_f is defined as the volume surrounding the center of a particle in which no other particle centers are present, as illustrated

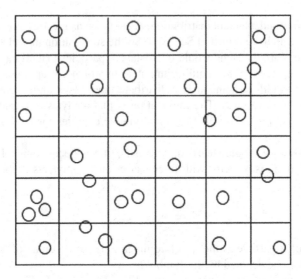

Fig. 3.1 Snapshot of $N = 36$ particles randomly distributed in a total volume V, which is divided in N cubes, each with a volume V/N. At any moment some cubes are empty and others contain one or more particle centers. On average each particle has a free volume $<v_f> = V/N$ available in which no other particle center is present. Note that finite particle volumes actually reduce the free volume such that $<v> = V/N$ only holds at sufficiently low concentrations

in Fig. 3.1. Due to particle motions this free volume will fluctuate in time, so we need to consider the *average* free volume per particle, denoted as $<v_f>$. According to the three assumptions discussed above particles are randomly distributed, non-interacting particles with a negligible volume. For N such particles in a total volume V, the mean free volume equals the inverse particle number density:

$$< v_f > = \frac{V}{N} = \frac{1}{\rho} \tag{3.6}$$

If the total particle volume would occupy a significant part of the system volume V, the mean-free volume would be smaller than given by (3.6) because less space is available for particle centers to move around. Note that for an ideal gas at pressure p and temperature T:

$$\frac{V}{N} = \frac{kT}{p} \tag{3.7}$$

Thus the average free volume is a constant for given p and T; this is nothing but a reformulation of Avogadro's principle stating that the molar volume of ideal particles is constant at given p, T.

Mean free path. Another way to estimate the free volume v_f is to look at the straight trajectory that particles traverse between two collisions. This trajectory is called the free path; Fig. 3.2 shows a particle that travels this free path with average

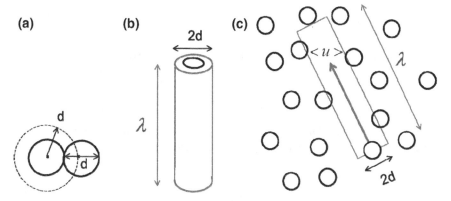

Fig. 3.2 a Particles are modeled as hard spheres with diameter d and collision cross section πd^2; a moving sphere sweeps a cylindrical volume with cross section πd^2 and collides with any sphere having its center located in this cylinder. **b** The length of the cylinder free from other centers is the mean free path λ. **c** A particle moves with average speed $<u>$ through a swarm of fixed neighbor spheres with a mean free path as indicated. The blue rectangle is the side view of a cylinder with length λ and cross section πd^2; the cylinder volume equals on average the free volume $<v_f>$

speed $<u>$ in a cloud of fixed neighbor spheres. The particle voyages unhindered until it collides with one of its static neighbors. The corresponding free volume is:

$$< v_f > = \lambda \pi d^2 \qquad (3.8)$$

Here λ is the average value of the free path, the mean free path, and πd^2 is the collision cross-section, see also Fig. 3.2b. Thus the average free volume $<v_f>$ can be eliminated from (3.6) and (3.8) to obtain:

$$\frac{1}{\rho} = \lambda \pi d^2, \qquad (3.9)$$

such that the mean free path is given by:

$$\lambda = \frac{1}{\rho \pi d^2} \qquad (3.10)$$

This is the mean free path for one particle moving with average speed $<u>$ in a swarm of *fixed* neighbors, see Fig. 3.2. In reality these neighbors are also in motion so the relative speed of particles with respect to each other is actually larger than $<u>$. This enhances the rate at which particles collide and, consequently, reduces the mean free path. This reduction turns out to be[2] a factor of $\sqrt{2}$:

---

[2]See f.e. W. J. Moore, *Physical Chemistry*, Longman, London, fifth ed. (1972), pp. 148–150.

$$\lambda = \frac{1}{\sqrt{2}\rho\pi d^2} \tag{3.11}$$

Using the ideal gas law $p = \rho kT$ we find the alternative expression

$$\lambda = \frac{kT}{\sqrt{2}\pi d^2 p} \tag{3.12}$$

So all we need is the diameter d of particles to calculate the mean free path in a gas at given pressure and temperature. The volume fraction ϕ of particles with diameter d is defined as:

$$\phi = \rho\left(\frac{\pi}{6}\right)d^3 \tag{3.13}$$

Combining (3.12) and (3.13) we find the simple expression:

$$\frac{\lambda}{d} = \frac{1}{\sqrt{2}\,6\phi} \tag{3.14}$$

For example, in a dilute gas of colloidal spheres with volume fraction $\phi \sim 0.001$, the colloids travel unhindered over distances $\lambda \sim 118$ times their own diameter d.

Collision frequencies. The result for the mean free path λ in (3.12) directly leads to an expression for the frequency z at which particles collide with one target particle. Again particles are modelled as hard spheres with diameter d. Realizing that the time t_B that elapses between two collisions equals $1/z$, the mean free path follows from:

$$\lambda = <u> t_B = <u> \frac{1}{z}, \tag{3.15}$$

where $<u>$ is the average speed of the particles. From (3.12) and (3.15) we can eliminate λ to obtain the collision frequency z of particles on one selected target:

$$z = <u> \sqrt{2}\,\rho\pi d^2 = <u> \sqrt{2}\,\pi d^2 \frac{p}{kT} \tag{3.16}$$

Also here we get a compact result in the terms of the particle volume fraction ϕ from (3.13):

$$z = \frac{<u>}{d}\sqrt{2}\,6\phi \tag{3.17}$$

This result allows a quick estimate of the order of magnitude of z. Taking, for example, molecules with diameter $d \sim 1$ Å that travel at the speed of sound ($u \sim 330$ m/s), the collision frequency is approximately:

$$z \approx 3 \times 10^{13} \phi\, s^{-1} \tag{3.18}$$

Thus even for low gas volume fractions gas collision frequencies have staggering values; for example, for $\varphi = 0.5 \times 10^{-4}$ a gas molecule collides in one second with about one *billion* other molecules.

Binary collisions. The frequency z denotes the collisions on one chosen particle; since there are in total N particles in a volume V, the total number Z_B of binary collisions, per second per volume equals:

$$Z_B = \frac{1}{2}z\frac{N}{V} = \frac{1}{2}z\rho = \frac{\pi d^2 <u>}{\sqrt{2}}\rho^2 \tag{3.19}$$

Here ρ is the particle's number density; the factor ½ is included to avoid double counting of collisions. Z_B provides the maximal rate for a reaction between gas particles, when every collision produces a dimer.

3.3 Pressure from Ideal Thermal Particles

The energy of ideal particles only comprises the kinetic energy that is stored in the their thermal motion; the same holds for ideal solute particles in a solution. The thermal motion of particles generates a pressure p and we wish to assess how pressure p is related to masses and velocities of the particles involved. An educated guess for this relation, based on dimensions, is as follows.

A dimensional conjecture. A pressure has the unit of $N/m^2 = Nm/m^3 = J/m^3$ so pressure is an energy in joule (J) per volume. Ideal particles have only one type of energy, and that is the kinetic energy stored in their translational motions. Thus N particles have a total kinetic energy $N<E_{kin}>$, where $<E_{kin}>$ is the average kinetic energy per particle. This total kinetic energy is present in a volume V so for the pressure to have the correct dimension we expect that:

$$p \propto \rho < E_{kin} >; \ \rho = \frac{N}{V}, \tag{3.20}$$

where ρ is the particle number density; the proportionality sign $\propto$ means 'is apart from a constant equal to'. The kinetic energy of a particle with mass m and speed u equals $E_{kin} = (1/2)mu^2$ so the proportionality (3.20) can also be written as:

$$p \propto \rho m < u^2 > \tag{3.21}$$

To find the missing constant of proportionality in (3.21) we will evaluate pressure p in a more rigorous fashion, starting from Newton's second law and the momentum transport by ideal particles.[3]

---

[3]The following derivation of ideal particle pressure is based on the treatment by James Clerk Maxwell (1831–1879) in his *Theory of Heat* (Longmans, London 1888); reprinted by Dover, Mineola, 2001.

Newton's second law. Newton's first law states that, due to its *inertia*, a particle remains at constant velocity unless acted upon by a net external force. For the particle velocity to change, either in magnitude or direction, a force F is required that according to Newton's second law is given by:

$$\vec{F} = m\frac{d\vec{v}}{dt} \tag{3.22}$$

The product of particles mass m and velocity $\vec{v}$ is the *momentum* $\vec{P}$ of the particle so Newton's second law can also be written as:

$$\vec{F} = \frac{d\vec{P}}{dt}; \quad \vec{P} = m\vec{v} \tag{3.23}$$

Note that this is a vector equation: both changes in magnitude and direction of the particle moment require a force. Below we will need the average momentum change in a time interval Δt, associated with a time-average force

$$< \vec{F} >_t = \frac{\Delta \vec{P}}{\Delta t} \tag{3.24}$$

Mixture of ideal particles. Particles in a gas travel at different speeds, according to the equilibrium distribution treated in Sect. 3.4. Suppose we give label $j=1$ to particles that have velocities with an x-component equal or very close to $v_{x,1}$, and do the same for groups of particles with labels $j=2$, 3, 4 ... Q, thereby dividing the particles into a mixture of Q components. We will first calculate the partial pressure exerted by one of these components, and then evaluate the total pressure by summation of the partial pressures of all Q components.

Momentum exchange. Imagine a suspended open wire with area O, as sketched[4] in Fig. 3.3. Particles travel through the area O but since in equilibrium there is no net particle transport, every particle that crosses the wire from L to R, is balanced by another particle that crosses from R to L with on average the same, but opposite x-component of the velocity. The momentum exchange associated with one particle with label j going from L to R and one j-particle going in reverse is

$$\Delta P_{x,j} = mv_{x,j} - (-mv_{x,j}) = 2mv_{x,j} \tag{3.25}$$

If this momentum exchange lasts Δt seconds, the average force in this time interval is, according to Newton's second law (3.24):

$$< F >_{t,L \to R} = \frac{2mv_{x,j}}{\Delta t} \tag{3.26}$$

---

[4]Figure 3.3 shows spherical molecules or colloids. However, the derivation of equation (3.35) does not presuppose that particles or solutes and holds for ideal particles of arbitrary shape.

Fig. 3.3 Two gas particles fly through an open wire with area O, respectively, from space L to R and *vice versa*. The net momentum transfer $2mv_x$ in a time interval Δt corresponds to an average force with magnitude $F_{L \to R} = 2mv_x/\Delta t$, from which we can calculate the gas pressure, as further explained in the text

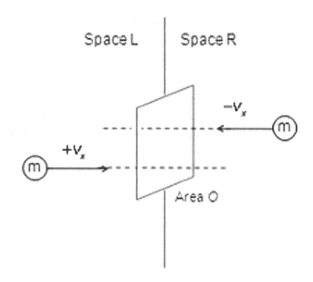

The number of j-particles that reach the wire within Δt seconds is

$$\frac{1}{2}v_{x_j}\Delta t \frac{N_j}{V}O \tag{3.27}$$

Here N_j is the number of j-particles in volume V; the factor $\tfrac{1}{2}$ takes into account that only 50% of j-particles in space L have an x-component of the velocity in the direction of the wire. The pressure exerted by the j-particles is the total force per unit area

$$p_{L \to R} = \frac{1}{O} \times \frac{1}{2}v_{x,j}\Delta t \frac{N_j}{V}O \times <F>_{t,L \to R} = \frac{N_j}{V}mv_{x,j}^2 \tag{3.28}$$

Windless equilibrium. According to the random-motion assumption from Sect. 3.1 particles move in all directions with equal probability so there are no net, macroscopic particle displacements. Thus the j-particles do not exert a net force on the area O, implying that in equilibrium the pressure $p_{L \to R}$ in (3.28) equals the opposite pressure $p_{R \to L}$. So we can drop the subscript $L \to R$ and write for the partial pressure exerted by the j-particles:

$$p_j = \frac{N_j}{V}mv_{x,j}^2 \tag{3.29}$$

Since we are dealing with ideal particles the total gas pressure p of the gas mixture equals, according to Dalton's law, the sum of all partial pressures p_j:

$$p = \sum_j p_j = \frac{m}{V}\sum_j N_j v_{x,j}^2, \tag{3.30}$$

where the summation runs over all groups $j=1, 2, 3, \ldots Q$ of gas particles. The average of the square of the velocity component v_x is defined by:

$$\langle v_x^2 \rangle = \frac{1}{N} \sum_j N_j v_{x,j}^2; \quad N = \sum_j N_j \tag{3.31}$$

Here N is the total number of particles, found by adding up all numbers of particles, N_j, of all components $j=1, 2, 3 \ldots, Q$. By combining Eqs. (3.30) and (3.31) we obtain for the total pressure:

$$p = \rho m < v_x^2 >; \quad \rho = \frac{N}{V} \tag{3.32}$$

The average of the square of the particle speed u is given by:

$$< u^2 > = < v_x^2 > + < v_y^2 > + < v_z^2 > \tag{3.33}$$

Again we exploit the random-motion assumption that in equilibrium, gas pressures are the same in all directions: no wind is blowing in an equilibrium gas. The implication is that

$$< v_x^2 > = < v_y^2 > = < v_z^2 > \tag{3.34}$$

Thus the pressure in (3.32) can be rewritten to:

$$p = \frac{1}{3} \rho m < u^2 >; \quad \rho = \frac{N}{V}, \tag{3.35}$$

which should be compared to our conjecture in Eq. (3.21). For the average kinetic energy we will derive in Sect. 3.4 that:

$$< E_{kin} > = \frac{1}{2} m < u^2 > = \frac{3}{2} kT, \tag{3.36}$$

which on substitution in (3.35) yields for the pressure:

$$p = \rho kT \tag{3.37}$$

This is the pressure law for ideal, non-interacting thermal particles. For particles in a gas phase, (3.37) is usually referred to as the ideal gas law; for particles that are solutes in a solution, (3.37) is commonly written as

$$\pi = \rho kT \tag{3.38}$$

Here the symbol π is used for the pressure to indicate that particles are immersed in a solvent rather than moving around in a gas vessel. The pressure exerted by solutes is also called the *osmotic pressure* and Eq. (3.38) is referred to as Van' t Hoff's law. We will return to the osmotic pressure of dilute solutions in Chap. 11, which comprises two thermodynamic alternatives for the kinetic derivation of Eq. (3.38) given above.

3.4 Velocity Distributions and Energy Equipartition

We have already encountered in Sect. 3.2 the average speed $<u>$ of gas particles that determines collision frequencies, and found in Sect. 3.3 that particle pressures depend on the average $<u^2>$ of the squares of particle speeds. We will now introduce the mathematical tools, required to evaluate averages such as $<u>$ and $<u^2>$.

Cars versus particles. For the determination of the average speed of cars on a high way, we sum the measured speeds of N cars, and divide the outcome by N to obtain the desired average. An analogous procedure for particles would be the determination of their average speed from the list of speeds of, say, one mole of particles. However, even for a tiny fraction of a mole such a list would be as colossal as it would be superfluous: an inventory of all individual velocity components contains very much more information than we actually need. To evaluate average quantities for large numbers of particles all we need is a *distribution function* for their velocity components.

Discrete distributions. To introduce particle distribution functions we first consider the *discrete* probability distribution of car speeds sketched in Fig. 3.4; here the speed-axis is divided in Q bins that each represent a certain range of speeds. For example, the bin 99–101 km/h contains the number of cars that have a speed u in the range 99–101 km/h. The average car speed follows from

$$\langle u \rangle = \frac{\sum_j n_j u_j}{\sum_j n_j} \tag{3.39}$$

Here n_j is the number of cars in bin j that, on average, have speed u_j. The summation sign, the upper case Greek letter Σ, denotes the summation over all bins labelled $j = 1, 2, 3 \ldots Q$. Note that the summation in the denominator in (3.39):

$$\sum_j n_j = n_1 + n_2 + n_3 + \cdots n_Q = N \tag{3.40}$$

yields the total number, N, of cars. The probability p_j to randomly select a car with speed u_j equals the number of cars in bin j divided by the total number of cars:

$$p_j = \frac{n_j}{\sum_j n_j} = \frac{n_j}{N} \tag{3.41}$$

Combination of (3.41) and (3.39) leads to the average car speed in the form:

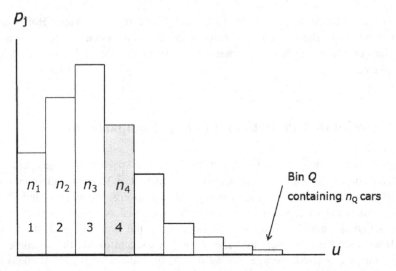

Fig. 3.4 Discontinuous probability distribution of car speeds; p_j is the probability to find a car in a bin with label $j = 1, 2, 3 \ldots Q$, containing n_j cars with average speed u_j. For example, p_4 is the fraction n_4/N of the total number N of cars in bin 4. The probability p_j is normalized, meaning that the sum of all probabilities equals one

$$\langle u \rangle = \sum_j p_j u_j \qquad (3.42)$$

Thus an average of a quantity follows from adding up all its values multiplied by their probability to occur. For example, the average of the square of car speeds is:

$$\langle u^2 \rangle = \sum_j p_j u_j^2 \qquad (3.43)$$

An alternative measure for average speed is the *root-mean-square* (index 'rms') speed, defined as:

$$u_{\mathrm{rms}} = \langle u^2 \rangle^{1/2} \qquad (3.44)$$

Returning back to the probability p_i in (3.41) we note that for an arbitrarily chosen car to have a (any) speed, that probability must be one. This requirement implies that the sum of all probabilities p_j equals one, which is indeed the case:

$$\sum_j p_j = \frac{\sum_j n_j}{N} = \frac{N}{N} = 1 \qquad (3.45)$$

A probability whose sum of all its values equals one, is called a *normalized* probability.

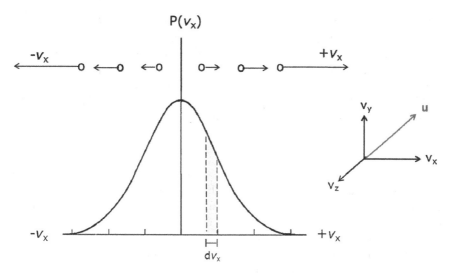

Fig. 3.5 Continuous distribution of components v_x of particle velocities. On the x-axis the magnitude of v_x runs from plus infinity to minus infinity; on the y-axis are the values of the distribution function $P(v_x)$. The shaded area is the probability $P(v_x)dv_x$ to find a velocity component in the interval between v_x and $v_x + dv_x$: the shaded area is the fraction of particle velocities with an x-component between v_x and $v_x + dv_x$ provided the distribution function is normalized, meaning that the total area under the curve equals one

Continuous distributions. For the case of velocities of astronomic numbers of particles, the bins as in Fig. 3.4 would be very narrow, in fact close to infinitesimally narrow, such that the distribution becomes *continuous*. Figure 3.5 shows a sketch of such a distribution for the x-component of particle velocities. In the discontinuous distribution of Fig. 3.4 the y-axis represents the probability p_j—defined in (3.41)—to find a car with speed u_j. In Fig. 3.5 the y-axis is the distribution function $P(v_x)$; its definition is that $P(v_x)dv_x$ equals the probability to find a velocity component between v_x and $v_x + dv_x$. Thus the average value of $(v_x)^2$ follows from:

$$< v_x^2 > = \int\limits_{-\infty}^{+\infty} P(v_x)v_x^2 dv_x \qquad (3.46)$$

As for the discrete probability p_j in (3.41) the continuous probability $P(v_x)dv_x$ must also fulfill the requirement of normalization. Instead of the discrete summation in (3.45) normalization is now the integral:

$$\int\limits_{-\infty}^{+\infty} P(v_x)dv_x = 1 \qquad (3.47)$$

Note the difference between the dimensionless probabilities p_i, defined in (3.41) for a discrete distribution and the distribution function $P(v_x)$: the latter is not a probability and has the dimension of reciprocal speed. A distribution function is also referred to as a probability *density* which turns into a numerical probability by multiplying with an infinitesimal interval of the variable in question (here dv_x).

Moments of a distribution. Suppose y is the distributed quantity of interest, with a normalized probability density $P(y)$. Then the average of y^m is given by:

$$< y^m > = \int y^m P(y) dy \tag{3.48}$$

Here $<y^m>$ is called the m-th moment of the distribution in y. Thus, the average $<v_x^2>$ in (3.46) represents the second moment of the distribution in velocity components v_x. Having identified a method to compute moments of a distribution, the question is now which distribution function $P(y)$ has to be substituted in (3.46). The answer is that an equilibrium probability density distribution for particle properties such as particle speeds and spatial positions is always a *Boltzmann distribution*.

The Boltzmann factor. In Chap. 2 (Exercise 2.2) we already encountered the equilibrium concentration profile resulting from the competition between Brownian motion and gravity, a profile that is further analyzed in Sect. 10.5. For particles with mass m the equilibrium concentration profile is

$$\rho(h) = \rho_0 \exp[-\frac{U_{pot}}{kT}]; \ U_{pot} = mgh \tag{3.49}$$

Here g is the gravitational acceleration and ρ_0 is the particle concentration at $h=0$; U_{pot} is the potential energy of a particle at height h. Since the probability $P(h)$ to find a particle at height h must be proportional to the concentration $\rho(h)$, we can infer from (3.49) the proportionality

$$P(h) \propto \exp[-\frac{U_{pot}}{kT}] \tag{3.50}$$

This is the exponential Boltzmann factor for particles in the gravity field. More generally we can write for the Boltzmann factor:

$$p(\varepsilon_j) \propto \exp[\frac{-\varepsilon_j}{kT}] \tag{3.51}$$

Here $p(\varepsilon_j)$ is the probability of finding a particle in a state j with energy ε_j, a probability which, for given energies, only depends on the absolute temperature T. Energy ε includes, among many others, the potential energy in the barometric height profile in (3.50), the discrete energy levels for particle rotations and vibrations and the continuous distributions of translational kinetic energies. Equation (3.51) can be turned into an equality by the normalization of the probability $p(\varepsilon_j)$ via:

$$p(\varepsilon_j) = \frac{\exp[-\varepsilon_j/kT]}{\sum_j \exp[-\varepsilon_j/kT]} \tag{3.52}$$

Here the summation is carried out over all energy levels (or quantum states), labelled $j = 1, 2, 3, \ldots$. Note that we deal here with a discrete distribution of energies; only if the spacing between energy levels is very small the distribution of energies and particles approaches a continuous one. For translational kinetic energies level spacings can be neglected, so the kinetic energy of particles and colloids can be described by a continuous distribution function, as the one shown in Fig. 3.5.

One-dimensional velocity distributions. For the continuous distribution of velocity components v_x the relevant energy in the Boltzmann factor is a particle's kinetic energy so the probability distribution function for velocity components v_x is given by:

$$P(v_x) = C \exp[-E_{kin,x}/kT] = C \exp[-mv_x^2/2kT] \tag{3.53}$$

Here $E_{kin,x}$ is the contribution of the x-component of the velocity to the kinetic energy of a particle. The constant C follows from the normalization requirement

$$C \int_{-\infty}^{+\infty} \exp[-mv_x^2/2kT]dv_x = 1 \tag{3.54}$$

Employing the Gaussian integral (see Appendix A)

$$\int_{-\infty}^{+\infty} e^{-ay^2}dy = \sqrt{\frac{\pi}{a}}, \tag{3.55}$$

the integral in (3.54) can be easily solved, with the result that for the probability $P(v_x)dv_x$ to be properly normalized, the constant must equal:

$$C = \left(\frac{m}{2\pi kT}\right)^{1/2} \tag{3.56}$$

So the normalized probability density for v_x reads

$$P(v_x) = \left(\frac{m}{2\pi kT}\right)^{1/2} \exp[-mv_x^2/2kT] \tag{3.57}$$

The rms-speed. Using the probability density in (3.57) we can now compute the average of the square of velocity components v_x by evaluating the following integral:

$$< v_x^2 > = \left(\frac{m}{2\pi kT}\right)^{1/2} \int_{-\infty}^{+\infty} \exp[-mv_x^2/2kT]\, v_x^2 dv_x \tag{3.58}$$

Rewriting this integral to:

$$< v_x^2 > = \left(\frac{a}{\pi}\right)^{1/2} \int\limits_{-\infty}^{+\infty} e^{-av_x^2} v_x^2 dv_x ; \quad a = \frac{m}{2kT}, \tag{3.59}$$

we can employ the Gaussian integral (3.55) to obtain[5]:

$$< v_x^2 > = \frac{kT}{m} \tag{3.60}$$

As a consequence of the random-motion assumption from Sect. 3.1 (averages are the same in all directions) the outcome for the y and z-velocity components must also equal kT/m. Thus the average of the squared speed is:

$$< u^2 > = < v_x^2 > + < v_y^2 > + < v_z^2 > = \frac{3kT}{m} \tag{3.61}$$

As a measure for the average particle speed we can employ the rms-speed introduced in Eq. (3.44); using the result for $<u^2>$ in (3.61) we obtain:

$$u_{\text{rms}} = \langle u^2 \rangle^{1/2} = \left(\frac{3kT}{m}\right)^{1/2} = \left(\frac{3R_g T}{M}\right)^{1/2} \tag{3.62}$$

Here m is the mass of a particle and M is the molar mass. Gas particles in the atmosphere race around at supersonic speeds; nitrogen particles at room temperature have according to (3.62) an rms-speed of 515 m/s whereas the somewhat heavier CO_2 molecules still achieve a respectable speed of 411 m/s which exceeds the speed of sound of 330 m/s. Sound, incidentally, is transmitted by molecular motions but since sound only moves in one direction it goes slower than the average molecular speed.

The kinetic energy of a particle equals:

$$E_{\text{kin}} = \frac{1}{2} mu^2, \tag{3.63}$$

so from the rms-speed in (3.62) it follows that the average kinetic energy only depends on temperature:

$$< E_{\text{kin}} > = \frac{3}{2} kT \text{ (per particle)} ; \quad < E_{\text{kin}} > = \frac{3}{2} RT \text{ (per mole)} \tag{3.64}$$

---

[5]$< v_x^2 > = -\left(\frac{a}{\pi}\right)^{1/2} \frac{d}{da} \int\limits_{-\infty}^{+\infty} e^{-av_x^2} dv_x = -\left(\frac{a}{\pi}\right)^{1/2} \frac{d}{da} \left(\frac{\pi}{a}\right)^{1/2} = \frac{1}{2a}$

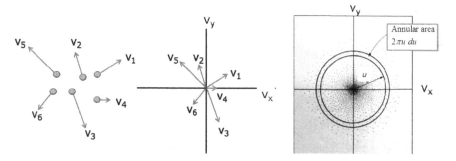

Fig. 3.6 Velocity vectors from a snapshot of particles (left) are transposed to the origin of a two-dimensional velocity space (middle). For very large numbers of particles the vector end points form a sphero-symmetrical cloud (right). Speed u is realized by all two-dimensional vectors (v_x, v_y) with their end-point in annular area $2\pi u du$. Three-dimensional velocity vectors (not shown here) have their end points in shell volume $4\pi u^2 du$

The 3-D Maxwell-Boltzmann (MB) distribution. Above we calculated $<u^2>$ which eventually leads to the expression for the rms-speed in (3.62). In this calculation we employed distribution functions for the velocity components $P(v_x) = P(v_y) = P(v_z)$, as given by Eq. (3.57). These are *one*-dimensional distribution functions. However, for the calculation of the average speed $<u>$ we need a *three*-dimensional distribution function. That is, we consider three-dimensional velocity vectors v and ask for the probability that the velocity has a certain magnitude u. To derive the corresponding distribution function $P(u)$ we proceed as follows.

Imagine we take a snapshot of a large number of randomly moving particles and draw for each particle its velocity vector, as shown for some selected particles in Fig. 3.6. Next these vectors are transposed such that they all start in the origin of a Cartesian space, as illustrated for the two-dimensional case in Fig. 3.6. The components of these vectors are statistically independent, uncorrelated quantities, in accordance with random-motion postulate from Sect. 3.1.

Now probabilities for independent quantities are multiplied: if the probability for coin 1 to fall on its head is $p(1) = \frac{1}{2}$, the probability that three independent coins all fall on their head is:

$$p(1) \times p(2) \times p(3) = \frac{1}{2} \times \frac{1}{2} \times \frac{1}{2} = \frac{1}{8} \qquad (3.65)$$

In complete analogy we have for the probability to find three velocity components between v_x and $v_x + dv_x$, v_y and $v_y + dv_y$, and v_z and $v_z + dv_z$, the product:

$$P(v_x)dv_x \times P(v_y)dv_y \times P(v_z)dv_z = C^3 \exp[-m(v_x^2 + v_y^2 + v_z^2)/2kT]dv_x dv_y dv_z$$

$$(3.66)$$

Here C is again the 'one-dimensional' normalization constant given by Eq. (3.56). From the theorem of Pythagoras

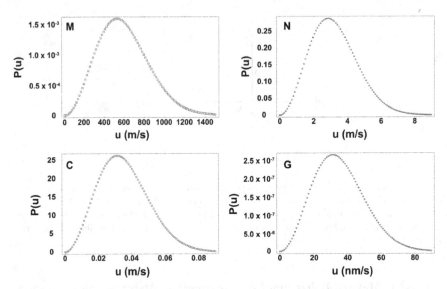

Fig. 3.7 Maxwell-Boltzmann distributions at $T = 298$ K for the Particle Quartet from Table 1.1. Probability density $P(u)$ is in m/s; speeds u are in m/s, except for the granular G-spheres (label **G**) where u is in nm/s. Molecular M-spheres (label **M**) travel on average at supersonic speeds, whereas nano-spheres (**N**) move at the speed of a quiet bike ride: 11 km/h. Colloidal spheres (**C**) roam around at several centi-meters per second and the very sluggish granular spheres (label **G**) need more than nine hours to cruise their own radius of one millimeter. Figure courtesy Samia Ouhadjji

$$u^2 = v_x^2 + v_y^2 + v_z^2, \tag{3.67}$$

it follows that a value for speed u is realized by any triplet (v_x, v_y, v_z) that satisfies (3.67), each Pythagorean triplet representing a vector that ends in a shell with radius u and thickness du, see also Fig. 3.6. The number of these vectors is proportional to the shell volume $4\pi u^2 du$ so the probability for a particle to have a speed u is the product of this shell volume and the Boltzmann factors in (3.66):

$$P(u)du = C^3 \exp[-mu^2/2kT]4\pi u^2 du; \quad C = \left(\frac{m}{2\pi kT}\right)^{1/2} \tag{3.68}$$

The probability density $P(u)$ is usually referred to as the Maxwell-Boltzmann (MB) distribution. Figure 3.7 depicts the widely different MB-distributions for the Particle Quartet from Table 1.1. Noteworthy is the strong speed reduction upon increase of molecular mass: molecular M-spheres move at supersonic speeds whereas the nano-spheres cruise at an average speed of only 11 km/h, the pace of a relaxed bike ride.

Having now the three-dimensional MB distribution (3.68) at our disposal we can compute the average particle speed:

$$< u > = \int_0^\infty P(u)udu = 4\pi \left(\frac{m}{2\pi kT}\right)^{3/2} \int_0^\infty \exp[-mu^2/2kT]u^3 du \qquad (3.69)$$

To evaluate this integral we employ[6]

$$\int_0^\infty e^{-ax^2}x^3 dx = \frac{1}{2a^2} \qquad (3.70)$$

Making the appropriate substitutions in (3.69), the outcome for the average particle speed turns out to be:

$$< u > = \left(\frac{8kT}{\pi m}\right)^{1/2} = \left(\frac{8R_g T}{\pi M}\right)^{1/2} \qquad (3.71)$$

The average speed $<u>$ is smaller than u_{rms} in (3.62) which is due to the inequality

$$\langle u^2 \rangle \geq \langle u \rangle^2 \qquad (3.72)$$

In words: the average of squared (speed) values always exceeds the square of the average; this is true for *any* distribution as shown in Appendix A. The average and rms-speed differ not only numerically, but also in their application: $<u>$ is employed for time-dependent processes such as collision frequencies (Sect. 3.1); the second moment $<u^2>$ of the speed distribution is needed to calculate features that involve kinetic energies, such as the pressure exerted by particles (Sect. 3.1) and the energies involved in collisions.

Universal Maxwell-Boltzmann distribution. The MB-distribution (3.68) shifts to larger speeds upon increasing the temperature—which is obvious as the average kinetic energies of particles increases. Augmenting particle mass is one way of shrinking the MB-distribution to smaller speeds, as illustrated in Fig. 3.7 showing the wide variation of MB-distributions for the Particle Quartet from Table 1.1. Interestingly, underlying all MB-distributions is one universal distribution which emerges as follows. The most probable speed u_{max} is the speed at which a MB-distribution has a maximum, and is given by (Exercise 3.6):

$$u_{max} = \left(\frac{2kT}{m}\right)^{1/2} \qquad (3.73)$$

---

[6] $\int_0^\infty e^{-ax^2}x^3 dx = -\frac{d}{da}\int_0^\infty e^{-ax^2}x dx \overset{x^2=z}{=} -\frac{1}{2}\frac{d}{da}\int_0^\infty e^{-az}dz = -\frac{1}{2}\frac{d}{da}\left(\frac{1}{a}\right) = \frac{1}{2a^2}$

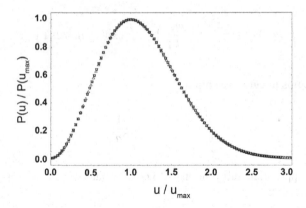

Fig. 3.8 When for the widely different distributions from Fig. 3.7, probability densities $P(u)$ and speeds u are scaled on their maximal value, all distributions collapse on one curve in accordance with Eq. (3.75)

Substitution of u_{max} in (3.68) yields:

$$P(u_{max})du_{max} = C^3 e^{-1} 4\pi u_{max}^2 du_{max}, \tag{3.74}$$

which can be combined with (3.68) to obtain:

$$\frac{P(u)}{P(u_{max})} = \left(\frac{u}{u_{max}}\right)^2 \exp\left[1 - \left(\frac{u}{u_{max}}\right)^2\right] \tag{3.75}$$

Rewritten in this form, the MB-distribution has become independent of temperature and particle mass. An example of the collapse of MD-distributions on this universal distribution is shown for the Particle Quartet from Table 1.1 in Figs. 3.7 and 3.8.

Effusion and Graham's law. Equation (3.71) for the average molecular speed can be used to predict the rate at which a gas effuses, i.e. the rate at which a dilute gas escapes from a vessel through a small hole (Fig. 3.9) or a porous membrane. Thomas Graham (1805–1869) found experimentally[7] that effusion rates of molecules with masses m_A and m_B obey the following relation, now known as Graham's law:

$$\frac{\text{rate (A)}}{\text{rate (B)}} = \left(\frac{m_B}{m_A}\right)^{1/2} \tag{3.76}$$

---

[7]Graham found (3.76) in 1831; the reason underlying this relation, namely Eq. (3.71), was elucidated only thirty years later by Maxwell's kinetic theory.

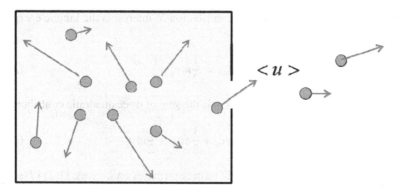

Fig. 3.9 Molecules escape from a dilute gas through a small open window at average speed $<u>$. The rate of this effusion is therefore inversely proportional to molecular mass, a proportionality known as Graham's law

Graham's law is a consequence of the relation between molecular velocities and molecular mass; from (3.71) we find for the ratio of the average speeds of molecules A and B:

$$\frac{<u>_A}{<u>_B} = \left(\frac{m_B}{m_A}\right)^{1/2} \tag{3.77}$$

Thus the effusion rate of gas molecules is proportional to their average speed which in turn is inversely proportional to molecular mass. For example, for oxygen molecules the effusion rate is $(32/2)^{1/2} = 4$ times lower than that of hydrogen molecules, allowing for separation of a hydrogen-oxygen gas mixture by a series of effusion chambers. Effusion is also applied to separate gas mixtures and fractionate gas mixtures containing isotopes.

For an aperture in the vessel with area A, the number Z_c of molecules that escape per second through the opening is:

$$Z_c = \rho A \int_0^\infty v_x P(v_x) dv_x = \rho A \left(\frac{kT}{2\pi m}\right)^{1/2} = \frac{1}{4}\rho A <u> \tag{3.78}$$

Here v_x is a velocity component in the positive x-direction towards the aperture; $P(v_x)$ is its distribution function from (3.57). Note that by choosing the integral limits 0 to ∞ we dismiss molecules that move away from the aperture.

Equipartition of energy. Boltzmann distributions, as we have seen in Sect. 3.2, can be employed to calculate average kinetic energies associated with velocity components in the x, y and z direction. There is a simple short-cut to find these average energies and that is to employ the equipartition theorem (ET) which states that for particles at thermal equilibrium each quadratic contribution to the energy equals on average $(1/2)\,kT$. Here the quadratic contribution of interest is the kinetic energy of a mass m moving in the x-direction:

$$E_{kin,x} = \frac{1}{2}mv_x^2 \tag{3.79}$$

The total kinetic energy of the mass m is the sum of three quadratic contributions:

$$E_{kin,x} = \frac{1}{2}mv_x^2 + \frac{1}{2}mv_y^2 + \frac{1}{2}mv_z^2 \tag{3.80}$$

According to the ET each quadratic term contributes on average $(1/2)\,kT$ so the total average kinetic energy per particle equals:

$$< E_{kin}> = \frac{3}{2}kT \tag{3.81}$$

This average kinetic energy is independent of particle mass m and only determined by temperature T. When at constant temperature and, hence, constant average kinetic energy, mass of particles increases, they move more slowly. Even large macroparticles and colloids of about one micron in radius still exhibit a significant thermal motion, manifesting itself as the Brownian motion that can be observed with an optical microscope.

The ET, it should be noted, is a theorem that is not always applicable; it only applies at temperatures that are high enough for many quantum energy levels to be populated. For molecular vibrational states, for example, the ET is unreliable as many energy levels are not accessible at room temperature. For the translational energies in (3.81), however, separations between energy levels are so small that many states are inhabited such that the ET is applicable to particle motions, from heavy colloids all the way down to small molecules.

3.5 Soft Matters

Thermal stability. Do thermal collisions carry enough energy to break chemical bonds? Apparently not for the gas molecules in the atmosphere: at room temperature colliding N_2 or O_2 molecules do not dissociate—for which we should be grateful because oxygen radicals are poisonous. The collisions-only assumption from Sect. 3.1 implies that for thermal collisions only kinetic energies have to be taken into account since the potential interaction energy of molecules is zero. For ideal molecules the average kinetic energy per mole is:

$$< E_{kin} > = \frac{3}{2}RT = 3.75\,\text{kJ mol}^{-1}; \quad T = 298\,\text{K} \tag{3.82}$$

This molar motional energy should be compared to typical dissociation energies of covalent bonds, given below in kJ/mol:

C–C   344
N–N   945
O–H   463
O–O   498

Clearly at room temperature thermal collisions are by far not capable of cleaving chemical bonds. However, these collisions can disrupt hydrogen bonds[8] that are considerably weaker than covalent bonds, corresponding to binding energies roughly in the range 10–20 kJ/mole. Dissociation of Van der Waals bonds[9], in turn, requires even less energy than hydrogen bonds. Water can exist as a liquid phase because of van der Waals and hydrogen bridges between water molecules. At higher temperature more bonds between water molecules are broken due to thermal collisions so more molecules migrate to the vapor phase; at the boiling point the liquid phase gradually disappears.

Soft forces. Water is an example of soft matter, a term denoting physical systems that are deformed or structurally changed by energies or stresses comparable to the thermal energy at room temperature. These systems also include colloids, polymers, foams, gels and many food products. Generally the characteristic length scale on which soft matter deforms and changes, is the mesoscopic colloidal length scale that is intermediate between the size of molecules and that of visible objects. In soft matter systems Brownian motion and thermal fluctuations are the agents for spontaneous change. And even if the forces involved are soft so to speak, the changes may be remarkable: colloids that crystalize into opal-like structures, space-filling gels that spontaneously shrink, and molecular motors that maneuver around in cells, to mention only three out of numerous examples.

Talking of cells: in biological systems and tissues the importance of thermal fluctuations is particularly significant. Interaction energies in biology are in the range of the thermal energy and it is by soft forces that life perseveres. The hope expressed in a Dutch poem[10] is that also in politics and society, soft forces will ultimately prevail. Here are its opening lines[11]:

The softer forces will no doubt prevail
in the end – this I hear as an intimate whisper
in me: were it mute all light would darken
all warmth within would stiffen

---

[8] Weak bonds between molecules resulting electrostatic attraction between a proton in one molecule and an electronegative atom in the other.

[9] Weak attractive force between electrically neutral molecules in close proximity caused by temporary attractions between electron-rich parts of one molecule and electron-poor parts of another.

[10] Written in 1918 by the Dutch poetess Henriette Roland Holst (1869–1952).

[11] My prose translation, see References for the original Dutch couplet.

Exercises

3.1 A bitter cold, extremely dilute hydrogen gas with a temperature of $T = 3$ K contains a number density of hydrogen atoms of 1 cm^{-3}. (a) Calculate the hydrogen pressure. (b) Estimate the mean free path for hydrogen atoms in the gas. (c) Estimate their collision frequency; how often does a hydrogen atom collide in one week and one year?

3.2 (a) Calculate the average free volume $<v_f>$ (in $\mathring{A}^3$) per water molecule in liquid water, assuming a mass density of $\delta = 1$ g cm^{-3} at $T = 298$ K.
    (b) In (a) we have not specified the pressure p that is exerted on the liquid water. Discuss whether this is a serious omission.
    (c) Calculate $<v_f>$ (again in $\mathring{A}^3$) for water particles in a water vapor at $T = 298$ K and a pressure of $p = 1$ bar. Assume the vapor obeys the ideal gas law.
    (d) Estimate the average distance between the particles in the vapor phase

3.3 Estimate the collision frequency on one particle in a gas with a volume fraction $\varphi = 0.5 \cdot 10^{-4}$. Assume that particles (with diameter 0.1 nm) move on average at the speed of sound.

3.4 Calculate the mean free path in a gas for the Particle Quartet from Table 1.1, for a particle volume fraction of one per cent.

3.5 Given are 2 cars with speed $u = 80$ km/h, 4 cars at $u = 100$ km/h and 10 cars at $u = 140$ km/h. (a) Calculate $<u>$. (b) Calculate the rms-speed of the cars. (c) Comment on any difference between the answers in (a) and (b).

3.6 Calculate the average molar translational kinetic energy at 300 K.

3.7 From the distribution function for particle speeds derived in this Chapter, show that the most probable speed is given by:

$$u_{max} = \left(\frac{2kT}{m}\right)^{1/2}$$

3.8 Derive a formula for the average square of the velocity x-component $<v_x^2>$ using the appropriate Boltzmann distribution. How large is $<v_x>$ and why?

3.9 Verify that

$$C = \left(\frac{m}{2\pi kT}\right)^{1/2}$$

for the distribution in Eq. (3.56) of this Chapter is correct.

3.10 Solve integrals (3.58) and (3.69) of Chap. 3.

3.11 Verify that the Gaussian distribution (see Appendix A) for the variable s

$$G(s) = \frac{1}{\sigma\sqrt{2\pi}} \exp[-\frac{1}{2}\left(\frac{s - <s>}{\sigma}\right)^2]; \quad \sigma^2 = <s^2> - <s>^2,$$

leads to the Maxwell-Boltzmann distribution for the velocity components v_x.

References

The standard work on the development of kinetic theory: S.G. Brush, *The kind of motion we call heat. A history of the kinetic theory of gases in the 19th century* (Amsterdam, North-Holland, 1994).

For a more in-depth treatment of kinetic theory see: J. Jeans, *An Introduction to the Kinetic Theory of Gases* (Cambridge University Press, 1940).

For an enthusiastic introduction into the field of soft matter, read: R. Piazza, *Soft Matter; the stuff that dreams are made of* (Springer Netherlands, 2011).

A lucid textbook, not only on gases is D. Tabor, *Gases, liquids and solids and other states of matter* (Cambridge University Press, third ed. 1991).

The first couplet of Henriette Roland Holst's poem in the Dutch original:
"De zachte krachten zullen zeker winnen
in't eind – dit hoor ik als een innig fluisteren
in mij; zo't zweeg zou alle licht verduistren
alle warmte zou verstarren van binnen".

Chapter 4
A Tale of Ten Time Scales

The three assumptions underlying kinetic theory mentioned at the start of Chap. 3 not only relate to kinetics of a molecular gas, or a colloidal mist of droplets in air, but equally apply to colloids that perform Brownian motion in a solvent. The colloids diffuse erratically around (assumption 1) at low concentrations such that they jointly occupy only a small volume fraction (assumption 3) and the effect of interactions is insignificant (assumption 2).

The motion of colloids and solvent molecules are similar in the sense that, according to the equipartition principle, they both carry the same average kinetic energy. An important difference, however, arises with respect to time scales: for solvent molecules, a colloidal particle is an extremely sluggish object, whereas the colloid experiences a dense swarm of molecules colliding with it at extremely high frequency. The wide range of time scales that confronts solute particles in a solution can be separated in two categories: one that relates to the inertial mass of colloids and one category that comprises diffusive times where, as we shall see, colloids have lost memory of their mass. This division follows from the distinction between Brownian and ballistic motion.

4.1 Brownian Versus Ballistic Motion

Buys Ballot's objection. One important result from the kinetic theory in Chap. 3 was the calculation of the average molecular speed:

$$< u >= \left(\frac{8kT}{\pi m} \right)^{1/2} = \left(\frac{8 R_g T}{\pi M} \right)^{1/2} \tag{4.1}$$

Here m is the mass of a molecule and M is its molar mass. We recall the implication of (4.1) that small gas molecules rush around at supersonic speeds. The Dutch

© Springer Nature Switzerland AG 2018
A. P. Philipse, *Brownian Motion*, Undergraduate Lecture Notes in Physics,
https://doi.org/10.1007/978-3-319-98053-9_4

mathematician and meteorologist Buys Ballot[1] (1817–1890) objected—actually not unreasonably—that this implication is at odds with everyday observations. If we open a bottle of perfume in the corner of a room, a seductive scent should be noticeable in the whole room within a split of a second, if gas molecules indeed migrate at the speed of sound. And if a stove would emit the lethal carbon monoxide, any attempt to escape would be futile: you cannot outrun the speed of sound! Buys Ballot's objection would cut ice if molecules indeed always travel along an uninterrupted straight line. Which they do not: molecular velocities change directions at every collision—and in Chap. 3 we concluded that these collisions occur at staggering frequencies.

In other words, the straight line that perfume or CO molecules freely traverse is very much smaller than the dimension of a room, and actually equals the average mean free path λ from Chap. 3:

$$\lambda = \frac{kT}{\sqrt{2}\,\pi d^2 p} \tag{4.2}$$

Here p is the pressure in a gas of (ideal) molecules with diameter d. To counteract Buys Ballot's objection with some numbers we note that at room temperature the average speed of a CO_2 molecule is 411 m/s. However, due to the many collisions the net displacement of a CO_2 molecule in the atmosphere is actually only 0.3 cm/s. Thus there is every opportunity to escape from poisonous gases—and, if you like, from perfume particles.

Ballistic motion. Unhindered molecular motion is also called *ballistic* motion; in accordance with Newton's second law the molecule's velocity will not change, neither in magnitude nor direction, in absence of any force on the molecule. The distance s travelled by ballistic motion of a molecule at average speed in time t is:

$$s = <u> t, \tag{4.3}$$

with $<u>$ given by (4.1). The sequence of molecular steps of magnitude λ that almost continuously change direction due to collisions is referred to as *diffusion*. The magnitude of the displacement r of a molecule by diffusion is given by:

$$\langle r^2 \rangle = 6Dt \tag{4.4}$$

Here $<r^2>$ is the average of the square of the displacement r in time t and D is the diffusion coefficient of the molecule. The distance that can be extracted from (4.4) is the root-mean-square (rms) value of the displacement r:

$$r_{\mathrm{rms}} = \langle r^2 \rangle^{1/2} = (6Dt)^{1/2} \tag{4.5}$$

---

[1] Known for Buys-Ballot's law which states that if a person on the Northern Hemisphere cycles against the wind, atmospheric pressure is higher at her left than at her right.

Note that for ballistic motion in (4.3) the distance s increases linearly with time t, whereas the rms-displacement by diffusion grows much slower, namely with the square root of time. This square-root dependence will be further addressed in Sect. 4.3.

4.2 Mass-Related Time Scales

The molecular collision time τ_C. The fastest process in a colloidal dispersion, relevant for Brownian motion, is the collision of solvent molecules with each other, and with a colloid. The average kinetic energy of particles with mass m and speed u equals

$$< E_{kin} > = < \frac{1}{2}mu^2 > = \frac{3}{2}kT \tag{4.6}$$

Since in a solvent the solvent molecules are closely packed together, molecules will collide when they travel a distance of about the molecular radius R. The time τ_C it takes to travers a distance R follows from;

$$R \sim \tau_C \sqrt{\langle u^2 \rangle} \tag{4.7}$$

So the collision time τ_C for molecules with radius R is of the order:

$$\tau_C \sim \frac{R}{\sqrt{kT/m}} \tag{4.8}$$

Here the symbol '$\sim$' should be read as "is approximately equal to". For molecular M-particles (Table 1.1) at room temperature we find from (4.6) that $<u^2>^{1/2} = 370 \, \text{m/s}$ and taking $R = 0.1 \, \text{nm}$ we obtain $\tau_C \sim 2.10^{-13} \, \text{s}$. Since the colloid is completely static on this time scale, τ_C is also the characteristic time for encounters between a colloid and its surrounding molecules. In other words, the molecules hit the colloid with a staggering frequency of order $1/\tau_C \sim 10^{13} \, \text{s}^{-1}$. This high frequency implies that on a time scale $t \gg \tau_C$ the colloid experiences a *continuous* fluid rather than a collection of discrete molecules. In such a fluid the motion of a colloid is damped by the Stokes friction and it is this viscous damping by which a colloid 'relaxes' its momentum.

The momentum relaxation time τ_{MR}. Suppose a colloidal particle is given an initial directed 'drift' velocity $\vec{v}_0$ and an initial momentum $\vec{P}_0 = m\vec{v}_0$ at time $t = 0$, see also Fig. 4.1. We ask for the time τ_{MR} it takes for this sphere to lose all its initial momentum due to viscous energy dissipation to the solvent. Assuming that the solvent behaves as a continuum, the viscous force on the sphere equals $f\vec{v}(t)$ where f is the Stokes friction factor. Newton's second law reads:

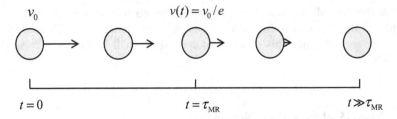

Fig. 4.1 The speed $v(t)$ of a sphere decays exponentially in time due to viscous kinetic energy dissipation. At the momentum relaxation time τ_{MR} the sphere speed has decreased by a factor e from its initial value v_0. Only a drift speed of the sphere imparted by an external force, will fully decay to zero. The thermal Brownian sphere will only partial decay as the average speed stays at the equilibrium value from Eq. (4.6)

$$\vec{F}_{tot} = \frac{d\vec{P}(t)}{dt}, \tag{4.9}$$

where $\vec{P}(t) = m\vec{v}(t)$ is the particle's momentum at time t. Since the total force on the sphere is $\vec{F}_{tot} = -f\vec{v}(t)$ we obtain the following differential equation for the velocity $\vec{v}(t)$ for a particle with constant mass m:

$$m\frac{d}{dt}\vec{v}(t) = -f\,\vec{v}(t)\ ,\quad \text{for } t \gg \tau_C\ , \tag{4.10}$$

from which we find that the instantaneous velocity $\vec{v}(t)$ and, consequently, the initial momentum decay as

$$\vec{v}(t) = \vec{v}_0 \exp\left[-\frac{ft}{m}\right] = \vec{v}_0 \exp\left[-\frac{t}{\tau_{MR}}\right] \tag{4.11}$$

The exponential decrease of the colloid's momentum (see also Fig. 4.1) is set by the decay time, also referred to as the momentum relaxation time:

$$\tau_{MR} = \frac{m}{f} = \frac{2}{9}\frac{\delta_p}{\eta}R^2, \tag{4.12}$$

for a sphere with mass $m = (4/3)\pi\delta_p R^3$ and friction factor $f = 6\pi\eta R$. For the standard colloidal C-sphere (see Table 1.1) in water we find that $\tau_{MR} = 5.10^{-9}$ s, which is four orders of magnitude longer than the molecular collision time τ_C: it was indeed justified to use in (4.10) the Stokes friction factor for a continuous, viscous fluid, see also Fig. 4.2. The distance $l(t)$ travelled by the sphere during the momentum relaxation process equals:

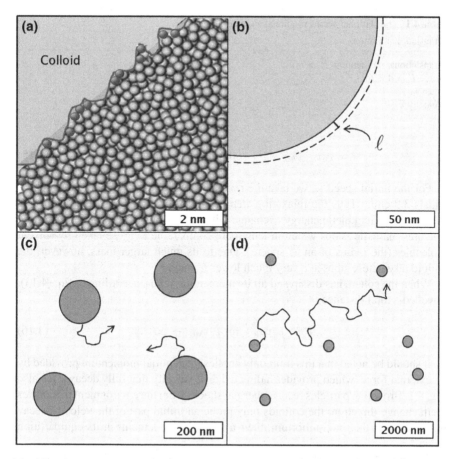

Fig. 4.2 a For solvent molecules, closely packed molecular M-spheres, a colloidal C-sphere is a bumpy plane with nano-scale surface irregularities. **b** C-spheres relax their momentum over a distance ℓ, on a time scale at which the solvent already behaves as a viscous continuum. After many relaxation steps C-spheres enter the diffusive regime where first configurations change in (**c**) followed at longer times by Brownian encounters in (**d**)

$$l(t) = \int_0^{t'} v(t)\mathrm{d}t = v_0\tau_{\mathrm{MR}}\big[1 - \exp(-t/\tau_{\mathrm{MR}})\big] \qquad (4.13)$$

For times much smaller than the momentum relaxation time we find from (4.13):

$$l(t) = v_0 t, \ \ \text{for } t \ll \tau_{\mathrm{MR}} \qquad (4.14)$$

This is the ballistic displacement by a sphere moving at uniform speed v_0. The typical momentum relaxation step ℓ is:

Table 4.1 Translational versus rotational angular motion

Translational motion	Angular motion
Translational momentum $\vec{P} = m\vec{v}$	$\vec{J} = I\vec{\Omega}$
Force $\vec{F} = \mathrm{d}\vec{P}/\mathrm{d}t$	Torque $\vec{T} = \mathrm{d}\vec{J}/\mathrm{d}t$
Mass m	Moment of inertia I
Velocity $\vec{v}$	Angular velocity $\vec{\Omega}$

$$\ell = v_0 \tau_{MR}[1 - e^{-1}] \approx 0.63 v_0 \tau_{MR}, \text{ for } t = \tau_{MR} \qquad (4.15)$$

For the initial speed v_0 we take the *rms*-speed which for the colloidal C-sphere equals 3.9 cm/s. Thus the relaxation step in (4.15) for the C-sphere is about $\ell \approx$ 0.1 nm. So in its kinetic energy exchange with the surrounding solvent, the colloid executes 'ballistic' steps which in length are comparable to those taken by solvent molecules (the radius of an M-sphere). Due to its much larger mass, however, the colloid takes these steps at a very much lower frequency.

When the sphere has dissipated all its momentum it has, according to Eq. (4.13), travelled a total distance of

$$l = v_0 \tau_{MR}, \text{ for } t/\tau_{MR} \rightarrow \infty \qquad (4.16)$$

It should be noted that this limit only applies to an initial momentum provided by an external force, which provides a directed drift velocity that fully decays, see also Fig. 4.1. Brownian particles receive thermal shocks from the environment that makes them change direction: the colloids only probe an initial part of the velocity decay in (4.11) such that in equilibrium, their average speed remains at its equipartition value.

The angular momentum relaxation time τ_{AR}. Suppose a sphere rotates at an angular velocity $\vec{\Omega}_0$ at $t = 0$. We ask for the time τ_{AR} it takes for the sphere to dissipate all its angular moment due to viscous friction by the solvent. Here it is helpful to note the analogy between translational and angular motion, summarized in Table 4.1.

Newton's second law for translational motion is given by (4.10); its equivalent for rotational motion is:

$$\frac{\mathrm{d}\vec{J}}{\mathrm{d}t} = -f_r \vec{\Omega}(t) \qquad (4.17)$$

where f_r is the rotational friction factor. The angular momentum is $\vec{J} = I\vec{\Omega}$, with I the sphere's moment of inertia. Integrating (4.17) yields for the angular particle velocity on time t:

$$\vec{\Omega}(t) = \vec{\Omega}_0 \exp[-f_r t/I] = \vec{\Omega}_0 \exp[-t/\tau_{AR}] \qquad (4.18)$$

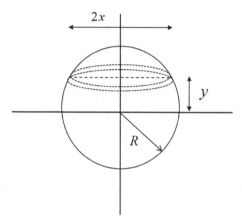

Fig. 4.3 Sketch for the calculation of the moment of inertia of a sphere in Sect. 4.2

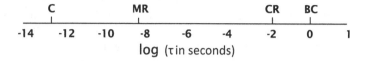

Fig. 4.4 Illustration of the broad range in times scales harbored by a dispersion of the Brownian C-spheres from Table 1.1. Indicated are the molecular collision time C, the momentum relaxation time MR, the time CR needed for a change in colloid configuration, and the typical time BC for spheres to encounter each other via Brownian motion in a dispersion with one volume percent of spheres

Here the relaxation time for the sphere's angular momentum is:

$$\tau_{AR} = \frac{I}{f_r} \tag{4.19}$$

The moment of a inertia I of a sphere is evaluated as follows. The mass of the slab in Fig. 4.3 with radius x is: (Fig. 4.4).

$$m_s = \pi x^2 \delta_p dy, \tag{4.20}$$

where δ is the mass density of the sphere. The moment of inertia of the slab about the axis Oy (O is the origin in Fig. 4.3) equals

$$I_s = \frac{1}{2} m_s x^2 = \frac{1}{2} \pi (R^2 - y^2)^2 \delta_p dy \tag{4.21}$$

So the moment of inertia of the whole sphere is found by the integration (Exercise 4.3):

$$I = \int_{-R}^{+R} I_s = \frac{2}{3} M R^2, \tag{4.22}$$

in which M is the total mass of the sphere. From (4.19) and (4.22) we find for the angular relaxation time, taking into account the rotational friction factor $f_r = 8\pi \eta R^3$:

$$\tau_{AR} = \frac{1}{15} \frac{\delta_p}{\eta} R^2 \tag{4.23}$$

Comparing τ_{AR} to the momentum relaxation time τ_{MR} in Eq. (4.12) we can conclude that sphere translations and sphere rotations decay on the same time scale. Thus on the diffusive time scale both translational and angular momenta have completely relaxed. In other words, the sum of all forces and the sum of all torques on a diffusing, Brownian particle are both zero.

The hydrodynamic decay time τ_{HD}. A moving colloid disturbs the surrounding fluid in two ways. First it causes pressure waves that travel at the speed of sound. Secondly the colloid motion initiates a shear wave, namely a flow pattern of fluid layers moving at different speeds (see for example Fig. 6.1). When a liquid layer moving in the x-direction contacts a slower layer, it transfers x-momentum to the slower layer. This momentum transfer is further discussed in Chap. 7, where it is concluded that the time τ_{HD} needed for momentum to travel via a shear wave a distance R is, in order of magnitude:

$$\tau_{HD} \sim \frac{\delta}{\eta} R^2 \sim \tau_{MR} \tag{4.24}$$

Here δ is the mass density of the fluid. This hydrodynamic decay time is comparable to the moment relaxation time τ_{MR} in Eq. (4.12) needed for a colloid to dissipate its (translational and angular) momentum; that makes sense because viscous dissipation is primarily losing momentum via shear waves. Also the propagation of pressure (sound) waves occurs on a time scale similar τ_{MR}. In water, for example, the velocity of sound is 1500 m/s so it takes about 10^{-9} s for a pressure disturbance to travel a distance of $R = 100$ nm.

Sedimentation time scales. The short time scales related to a colloid's inertial mass cannot be accessed by optical microscopy of Brownian motion—which takes place on the diffusive time scale to be discussed below. There is, however, another experiment by which we can observe the colloids in such a way that a momentum relaxation time can be measured. That is the observation of colloids settling under gravity in a sedimentation experiment. A boundary of sedimenting particles moves at speed u and leaves behind a clear solvent. The buoyant weight of a colloidal sphere is the product of buoyant mass Δm and the gravitational acceleration g. When the sphere has reached a stationary settling speed u, it experiences a friction force fu that balances the sphere's weight such that:

$$u = \frac{\Delta mg}{f} \tag{4.25}$$

Thus from a measured sedimentation speed we obtain the characteristic time τ_{MR}^{Δ}

$$\tau_{MR}^{\Delta} = \frac{u}{g} = \frac{\Delta m}{f} = \frac{2}{9} \frac{(\delta_p - \delta_{solv})}{\eta} R^2, \tag{4.26}$$

where δ_{solv} is the solvent mass density. By comparison with the momentum relaxation time τ_{MR} for mass m in (4.12) we see that (4.26) represents the momentum relaxation time for the smaller buoyant mass Δm. Another time scale in sedimentation is the time τ_{sed} it takes for a sphere to settle a distance equal to its own radius:

$$\tau_{sed} = R \frac{f}{\Delta mg} = \frac{R}{g} \frac{1}{\tau_{MR}^{\Delta}} \tag{4.27}$$

For nano-particles this sedimentation time is very long in comparison to the diffusion times scale discussed in Sect. 4.3. As a result particles remain under gravity homogeneously distributed in suspension. Employing an ultra-centrifuge the centrifugal acceleration may be orders of magnitude larger than g such that settling of nano-particles can be detected.

4.3   The Diffusive Regime

The time scales discussed above all derive from the mass of colloids or molecules. One of the intriguing aspects of Brownian motion is that in the course of time a colloid loses memory of its own mass. Below we will investigate the origin of this inertial oblivion that happens when the colloid enters the diffusive regime.

The diffusive or Brownian time scale $t \gg \tau_{MR}$. When a colloid has performed many moment-exchanging steps it enters the diffusive time regime $t \gg \tau_{MR}$, see also Fig. 4.5. Consider a sphere on a straight line that executes ballistic steps on the relaxation time scale τ_{MR}. The step length will be distributed; on the basis of (4.16) we consider an average displacement of order $\ell \sim v_0 \tau_{MR}$. The step frequency is typically $1/\tau_{MR}$ so if all steps would be in the same direction the net displacement x at time t equals:

$$x \sim \frac{l}{\tau_{mr}} t \sim v_0 t \ ; \ v_0 \sim \sqrt{3kT/m}; \tag{4.28}$$

This result represents uniform, ballistic motion at a constant speed v_0. The uniform displacement increases linearly in time and depends on the colloid mass. However, in the case of Brownian motion the momentum exchange with the solvent is a random process (Fig. 4.5): the sphere steps with equal probability either to the left (unit vector $-\hat{\delta}$) or to the right (unit vector $+\hat{\delta}$). Then the average displacement $<x>$ equals zero, by definition. For the square of the displacement vector we find on average for a large number $n = t/\tau_{MR} \gg 1$ of ballistic steps:

$$
< \vec{x} \cdot \vec{x} > = < \sum_{j=1}^{n} l\,\hat{\delta}_j \cdot \sum_{k=1}^{n} l\,\hat{\delta}_k >
$$

$$
= l^2 \frac{t}{\tau_{MR}} < \hat{\delta}_j.\hat{\delta}_j > + l^2 \sum_{j \neq k}^{n} \sum^{n} < \hat{\delta}_j.\hat{\delta}_k > = l^2 \frac{t}{\tau_{MR}} \qquad (4.29)
$$

The average of the double-summation of cross-terms $j \neq k$ vanishes because the summation produces a sequence of dot products of unit vectors equal to 1 or -1 with equal probability. Since

$$
l^2 \sim (v_0 \tau_{MR})^2 \sim \frac{kT\,\tau_{MR}^2}{m}, \qquad (4.30)
$$

we find a mean quadratic displacement that is proportional to:

$$
< x^2 > \sim \frac{kT}{m} \tau_{MR} t \sim \frac{kT}{f} t \quad , \text{ for } t \gg \tau_{MR} \qquad (4.31)
$$

Two conclusions here: (1) the colloid mass m has dropped out of the equation, and has no effect on diffusive displacements, and (2) the mean of the squared displacements growths linearly in time, in contrast to the ballistic motion in Eq. (4.28) for which the squared displacement increases *quadratically* in time. The difference is due to the circumstance that the ballistic steps all steps are all added, whereas for Brownian motion, steps are not only added but just as frequently subtracted. We recognize in the proportionality constant in (4.31) the translational diffusion coefficient $D = kT/f$. The correct factor multiplying time t in (4.31) is actually $2kT/f$, as we will show later in Chap. 6.

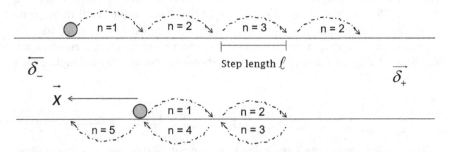

Fig. 4.5 Top: a sphere executes ballistic steps of length l in the same direction such that its displacement is proportional to time t. Bottom: a Brownian sphere steps with equal probability to the left or to the right which entails a net displacement that grows as the square root of t

The configurational relaxation time τ_{CR}. The position vectors that locate the centers of N Brownian particles are:

$$\vec{r_1} = (r_{1,x}, r_{1,y}, r_{1,z}); \; \vec{r_1} = (r_{2,x}, r_{2,y}, r_{2,z}); \dots \vec{r_N} = (r_{N,x}, r_{N,y}, r_{N,z}) \quad (4.32)$$

The N centers are positioned a in a 3-dimensional configuration space with Cartesian axes x, y, and z. How much time does it take for this configuration of colloids to change? In the mass-dominated, ballistic time regime the colloids make momentum relaxation steps, as a result of which the colloid configuration only changes on the sub-nanometer scale. More interesting are changes that—for micron size colloids could be observed under a microscope. This implies that colloids have entered the diffusive regime, and have diffused a distance comparable to their own radius R. Accordingly, a relaxation time τ_{CR} can be defined as the time needed for sphere centers to inscribe an area of order R^2 by diffusion. From Eq. (4.29) we obtain for this configurational relaxation time:

$$\tau_{CR} \sim \left(\frac{R}{l}\right)^2 \tau_{MR} \quad (4.33)$$

Clearly relaxation of particle configurations is extremely slow in comparison to moment relaxation. For the C-particle from Table 1.1 with $R = 100$ nm and $\ell = 0.1$ nm the difference is six orders of magnitude: $\tau_{CR} \sim 5.10^{-3}$ s versus $\tau_{MR} \sim 5.10^{-9}$ s. To put such a wide time span into human perspective: if it would take colloids one minute to relax their momentum, it would take more than one year before they have changed their configuration.

An alternative expression for τ_{CR} follows from (4.31):

$$\tau_{CR} \sim \frac{fR^2}{kT} \sim \frac{\eta R^3}{kT}, \quad (4.34)$$

This expression will return in our discussion of diffusion-controlled processes in Chap. 9.

Note that the time it takes for colloids to significantly change their configuration is very much longer than the time it takes for hydrodynamic disturbances (either sound or shear) to propagate. When we slightly displace a sphere in an arrangement of colloids, the flow field in the surrounding solvent almost instantaneously adapt itself. In other words, in the time region

$$\tau_{MR} \sim t \ll \tau_{CR} \quad (4.35)$$

colloids only experience each other (or a wall, or any other obstacle) via hydrodynamic flow fields ('hydrodynamic interactions'). Only on the timescale $t > \tau_{CR}$ the colloids encounter each other directly (Fig. 4.2) and experience the colloidal or 'direct' interactions.

Fig. 4.6 Cartoon accompanying the estimate of the Brownian collision time τ_{BC} from Eq. (4.37). A sphere inscribes by diffusion an area of $2Dt$ per second during which it collides with other spheres that have their centers (black dots) in a volume of order DRt

The rotational relaxation time τ_{RR}. When colloidal spheres have performed many angular steps in which they exchange angular momentum with the solvent, they enter the diffusive time regime $t \gg \tau_{AR}$. Initially the net angular displacement θ of the sphere is still insignificant. For θ to deviate substantially from its value $\theta = 0$ at $t = 0$ we have to wait at least a time.

$$\tau_{RR} = \frac{1}{D_r} \sim \frac{\eta R^3}{kT} \tag{4.36}$$

Here $D_r = kT/8\pi\eta R^3$ is the rotational diffusion coefficient that determines the decay of sphere orientations. Thus the time τ_{CR} in Eq. (4.34) taken by a sphere to significantly change its position coincides with the time τ_{RR} needed to substantially change its orientation.

The Brownian collision time τ_{BC}. We have already evaluated the collision time τ_C for molecules and it will come as no surprise that the collision time τ_{BC} for Brownian spheres will be *very* much longer. This collision time follows from Smoluchowski's theory for rapid coagulation to be discussed in Chap. 8; here we give a brief argument that leads to the same estimate for τ_{BC}.

Consider a tracer sphere (Fig. 4.6) with radius R diffusing in a dilute dispersion with low colloid number density ρ. The tracer sweeps by diffusion an area of the order of Dt square meters in t seconds. Since spheres collide at a center-to-center distance $2R$ the tracer scoops up in t seconds a 'collision volume' $2DRt$ in which it encounters about ρDRt other spheres. Therefore, the typical time between two such encounters is of the order:

$$\tau_{BC} \sim \frac{1}{\rho DR} \tag{4.37}$$

Substituting the sphere diffusion coefficient $D = kT/6\pi\eta R$ and the sphere volume fraction $\varphi = \rho(4/3)\pi R^3$ we can rewrite this to:

$$\tau_{BC} \sim \frac{\eta R^3}{kT\varphi} = \frac{\tau_{CR}}{\varphi}, \tag{4.38}$$

where τ_{CR} is the configurational relaxation time from Eq. (4.34). The R-dependence in (4.38) stems from the conversion of number density to volume fraction: if for a given φ we reduce the particle radius, the number density increases and particles collide at higher frequency. The time scale τ_{BC} determines the coagulation kinetics of colloids to which we will return in Chap. 9.

Characteristic time ratio's. To balance the effect of the erratic Brownian motion to the directed motion induced by an external force (gravity, liquid convection) it is convenient to evaluate the ratio of the time scales involved. For example, for colloids or nano-particles that settle under gravity, the relevant ratio is that of the configurational time scale to the sedimentation time scale:

$$\frac{\tau_{CR}}{\tau_{SED}} = \frac{4\pi R^4}{kT}(\delta_p - \delta_{SOLV}) \qquad (4.39)$$

It is seen that upon increase of the colloid radius, diffusive changes in colloid configuration slow down considerably in comparison to the rate of sedimentation. For a granular sphere the effect of Brownian motion during sedimentation can be disregarded.

The Péclet number. The competition between erratic Brownian motion and a directed, convective speed v can also be quantified by the Péclet number[2]:

$$Pe = \frac{Lv}{2D} \qquad (4.40)$$

Here L is a characteristic distance over which diffusion and convection takes place. Taking $L \sim R$ and substituting the sedimentation speed u from (4.25), the Péclet number equals the time ratio in (4.39).

Viscosity's dual role. On inspection of the mass-related time scales from Sect. 4.1 we observe that they are all inversely proportional to the solvent viscosity; the diffusive time scales, on the other hand, linearly increase with viscosity. Thus the solvent viscosity pulls mass-related and diffusive time scales apart. Consider, for example, the ratio between the configurational and momentum-relaxation time:

$$\frac{\tau_{CR}}{\tau_{MR}} = \frac{9R}{2\delta kT}\eta^2 \qquad (4.41)$$

When the viscosity rises the rate at which momenta relax increases but diffusion slows down due to the enhanced Stokes friction factor.

Measuring microscopic speeds. When colloids in a liquid dispersion are examined by optical microscopy, or probed by dynamic light scattering, the observation time τ_{OBS} is far beyond the momentum relaxation time τ_{MR}. Thus these techniques only provide diffusion coefficients but give no information on microscopic speeds of the suspended particles. However, for micron-size colloids suspended in a gas of low pressure, microscopic speeds can be measured, and the transition from ballistic

---

[2]Named after the French physicist Jean Claude Eugene Péclet (1793–1857).

to diffusive motion can be detected in the mean-square-displacements of a single colloidal sphere[3]. The sphere, incidentally, is held here in an optical trap such that radiation pressure prevents the colloid to settle under gravity.

Exercises

4.1   Calculate how far a marble with radius $r = 1$ cm diffuses in water in a century. ($T = 298$ K).

4.2   (a) Calculate using equipartition the *rms* velocity of a sphere with radius 100 nm and mass density $\delta_p = 1.5$ g cm^3 at $T = 298\ K$.
       (b) How large is the distance the sphere would traverse in one second with this velocity in uninterrupted linear motion?
       (c) How large is the *rms* displacement in one second in case of Brownian motion of the sphere? (Explain any difference with b).

4.3   Verify the derivation of the inertia moment of a sphere in (4.22).

4.4   Compute for all four particles in Table 1.1 the configurational relaxation time from (4.34).

4.5   (a) Estimate how many momentum relaxation steps a C-sphere has made within the time the sphere has diffused a distance equal to its radius. [See Eq. (4.33)].
       (b) What is the total distance covered by these relaxation steps?

4.6   Calculate the work w done by the friction force $f v(t)$ on a sphere with mass m during momentum relaxation. Interpret your finding.

References

For further discussion of ballistic versus diffusive motion see: P. N. Pusey, *Brownian Motion Goes Ballistic*, Science **332**, 802 (2011).

For a treatise of colloidal timescales in relation to dynamic light scattering see: P. N. Pusey and R. J. A. Tough in R. Pecora (ed.) *Dynamic Light Scattering; applications of Photon Correlation Spectroscopy* (Springer US, 1985).

---

[3]T. Lie et al., *Science* **328**, 1673 (2010).

Chapter 5
Continuity, Gradients and Fick's Diffusion Laws

Brownian motion is a sequence of random steps in positions or orientations of colloidal particles. Such a diffusive sequence can be described by a diffusion equation that quantifies how particle positions and orientations evolve in time. Diffusion belongs to the large class of transport phenomena, with other members such as the transport of heat or electricity, and also the viscous liquid flow treated in Chap. 7. Transport phenomena obey certain *conservation* laws, which stipulate that some quantity is conserved. For example, when colloids diffuse in a closed system, their total mass will remain constant. Such a conservation of mass, charge, energy or any other quantity is conveniently expressed in the language of vector calculus (see Appendix B) by a *continuity* equation.

Transport phenomena, of course, differ with respect to the substance that is displaced, and the type of force or gradient that sets the substance in motion. These distinctions are described by *constitutive* equations. One of them is Fick's first law for the diffusion flux, which we will apply in Sect. 5.4 to diffusion in a dilute gas.

5.1 The Continuity Equation

We start with the notion of a *flux density* which—in the case of diffusion—is the flow of particles through a unit area per second. Consider the flux density $\vec{j}$ of some property f, such as the concentration of molecules or colloids in a fluid, in which case the magnitude of $\vec{j}$ is the flux density of particles through a unit area per second. Suppose there are no 'sources' or 'sinks' for colloids in the fluid; in other words, no new colloids are being formed and no colloids disappear. Then the total number of colloids is conserved, which is formulated in vector notation as follows.

Consider a surface S enclosing a region V in a fluid with colloids (Fig. 5.1), with a normal unit vector $\vec{n}$ pointing outwards from a surface element dS. The flux density

© Springer Nature Switzerland AG 2018
A. P. Philipse, *Brownian Motion*, Undergraduate Lecture Notes in Physics,
https://doi.org/10.1007/978-3-319-98053-9_5

Fig. 5.1 Particle flux $\vec{j}$
through the surface of a
volume V

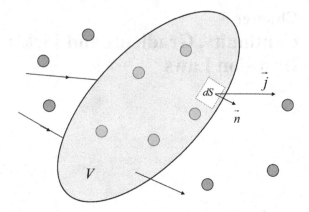

component along the normal is the dot product $\vec{j} \cdot \vec{n}$ so the net amount of f flowing
through the total surface S in unit time is the surface integral:

$$\int_S \left(\vec{j} \cdot \vec{n} \right) dS \tag{5.1}$$

Note the sign convention: in the case of a closed surface as in Fig. 5.1 the normal
is pointing in the positive, outward direction and the surface integral (5.1) is positive
if f is leaving the volume.

Suppose f is the local number density of colloids; *i.e.* the colloid concentration
in a small volume element dV at some position in the volume V in Fig. 5.1. Then at
any time t the total number of colloids in volume V is the volume integral:

$$\int_V f \, dV \tag{5.2}$$

The rate at which this total number of colloids changes in time is the derivative:

$$\frac{d}{dt} \int_V f \, dV = \int_V \frac{\partial f}{\partial t} dV, \tag{5.3}$$

where it is assumed that $\partial f / \partial t$ is continuous such that the derivative can be moved
under the integral sign. Conservation of the total number of colloids requires that the
number of colloids passing through the surface in Fig. 5.1 equals the change in the
number of colloids in the volume V:

$$\int_S \left(\vec{j} \cdot \vec{n} \right) dS = - \int_V \frac{\partial f}{\partial t} dV \tag{5.4}$$

According to the divergence theorem (see Appendix B) the surface integral in the LHS of (5.4) also equals:

$$\int_S \left(\vec{j} \cdot \vec{n}\right) dS = \int_V \vec{\nabla} \cdot \vec{j} \, dV \qquad (5.5)$$

The physical significance of this theorem for colloid transport is as follows. The divergence $\vec{\nabla} \cdot \vec{j}$ is the net colloid flow, per unit volume, out of a volume element. This volume element has a positive divergence. The outgoing colloids enter another volume element, thereby contributing to an opposite, negative divergence. Thus in the volume-integral in the RHS of Eq. (5.5) all divergences cancel, except for colloids leaving or entering the volume V through its surface, as quantified by the surface integral in the LHS of Eq. (5.5). From Eqs. (5.4) and (5.5) it follows that:

$$\int_V \left(\frac{\partial f}{\partial t} + \vec{\nabla} \cdot \vec{j}\right) dV = 0, \qquad (5.6)$$

The volume integral in Eq. (5.6) being zero does not necessarily imply a zero integrand. One could imagine a source inside V (integrand positive) which is exactly compensated by a sink (integrand negative). However, we already excluded the existence of sources and sinks inside V so the quantity f is conserved everywhere in V. Under this assumption the integrand in (5.6) is zero:

$$\frac{\partial f}{\partial t} + \vec{\nabla} \cdot \vec{j} = 0 \; , \qquad (5.7)$$

This is the continuity equation, a basic equation both in diffusion (Chap. 6) and hydro-dynamics (Chap. 7), with no other physical meaning than that f is a conserved quantity. We have found (5.7) taking as instance for f the local colloid concentration. However, the continuity equation is a completely general conservation law.

Incompressible fluids. Let us apply (5.7) to the mass flux $\vec{j} = \delta \vec{v}$ of a fluid with velocity $\vec{v}$ and mass density δ:

$$\frac{\partial \delta}{\partial t} + \vec{\nabla} \cdot (\delta \vec{v}) = 0, \qquad (5.8)$$

In many cases, for example a fluid such as water, the mass density has at each position the same constant, time-independent value. Then $d\delta/dt = 0$ such that (5.8) simplifies to:

$$\vec{\nabla} \cdot \vec{v} = 0, \quad \delta = \text{constant} \qquad (5.9)$$

This is the continuity equation for an incompressible fluid. As we will see later in Chap. 7 on hydrodynamics, the continuity Eq. (5.9) is an important constraint on

Table 5.1   Flux = transport property x gradient

Flux	Transport property	Gradient
Particles	Diffusivity	Particle density (Fick)
Charge	Conductivity	Electrical potential (Ohm)
Liquid	Permeability	Pressure (Darcy)
Momentum	Viscosity	Momentum density (Newton)
Energy	Heat conductivity	Temperature (Fourier)

the velocity field $\vec{v}$ around a colloid in a suspension, because the solvent usually *is* an incompressible liquid.

Stationary states. Equation (5.9) describes a *steady state* which by definition means that the distribution of the quantity f in (5.7) is time-independent. In a stationary flow of colloids in Fig. 5.1, for example, colloids enter and leave volume V at the same rate such that the colloid concentration f remains constant. Then $\partial f/\partial t = 0$ so from the continuity Eq. (5.7) it follows that the steady state automatically satisfies:

$$\vec{\nabla} \cdot \vec{j} = 0 \tag{5.10}$$

In the steady state the *divergence* of the flux is zero, which should not be confused with the thermodynamic equilibrium state in which the flux itself is zero. The fluxes in the steady state are due to irreversible processes (diffusion, viscous flow), which produce entropy, whereas in thermodynamic equilibrium (for example, as in the Maxwell-Boltzmann distribution from Chap. 3) no entropy producing transport processes can occur. One can also view equilibrium as the limiting case of a steady state in which all fluxes vanish. Examples of stationary states will be addressed in Sect. 5.3.

5.2   Constitutive Equations and Fick's Laws

The conservation Eq. (5.7) has two unknowns so to find the quantity f a second relation between f and its flux $\vec{j}$ is needed. Such a relation is a constitutive equation which specifies the transport problem and identifies the gradient that is responsible for the existence of the flux $\vec{j}$. An example is a concentration gradient of colloids which drives a net diffusive displacement of particles. The concept of a flux driven by a gradient of an intensive variable is quite general, as is illustrated by Table 5.1.

The momentum flux will be dealt with later in Chap. 7; liquid flow according to Darcy's law is briefly addressed in Sect. 8.1. Here we will continue with the formulation of Fick's diffusion laws.

Brownian motion is a random motion: colloids diffuse in all directions with equal probability. Thus there is no *net* displacement of particles in a homogeneous distribution with a constant concentration of colloids. A concentration gradient, however, induces a collective displacement of colloids, also referred to as gradient or *collective diffusion*.

Fick's first law. The diffusion flux $\vec{j}_d$ increases linearly with the concentration gradients $\vec{\nabla}\rho$, if that gradient is small enough. The diffusion coefficient D is defined as the ratio of the flux to the gradient that drives it:

$$D \equiv -\frac{\vec{j}_d}{\vec{\nabla}\rho}, \tag{5.11}$$

Here the symbol '$\equiv$' denotes a definition; ρ is the colloid concentration. Equation (5.11) is usually rewritten to

$$\vec{j}_d = -D\,\vec{\nabla}\rho, \tag{5.12}$$

a constitutive equation that goes by the name of *Fick's first law*—which is not so much a law then a definition of D. Fick's first law will be utilized later for the calculation the diffusion coefficient of a particle in a dilute gas in Sect. 5.4, and of a colloid in solution in Sect. 6.1.

We note here in passing that the collective diffusion coefficient D in Eq. (5.12) should, in principle, be distinguished from the *self-diffusion* constant for a single particle. The latter refers to the tortuous path of one colloid diffusing through a swarm of neighbor colloids. At infinite dilution, however, the magnitudes of the collective and the self-diffusion coefficient are the same. Since in this book we mostly neglect concentration effects, the coefficient D—unless stated otherwise—refers to the Stokes-Einstein coefficient of a single, free particle.

The diffusion equation. The number of colloids is conserved so the conservation law Eq. (5.7) yields:

$$\frac{\partial\rho}{\partial t} = -\vec{\nabla}\cdot\vec{j}_d, \tag{5.13}$$

which is also known as *Fick's second law*. If there is only a particle flux due to diffusion we can substitute Eq. (5.12) to obtain:

$$\frac{\partial\rho}{\partial t} = -\vec{\nabla}\cdot(-D\,\vec{\nabla}\rho) = D\,\nabla^2\rho, \tag{5.14}$$

assuming that the diffusion coefficient D is a constant (see Appendix B for further explanation of the vector notation). A solution of the diffusion Eq. (5.14) (Exercise 5.1) gives the number density $\rho(x,t)$ of the diffusing particles at every point in space at any time. Examples of number density profiles are the bell-shaped curves in Fig. 5.2, formed by particles diffusing away from a plate source.

Probability density. The solution of Fick's second law also provides the probability density $P(x,t)$ that entails the probability $P(x,t)dx$ to find a diffusing particle in the region between x and $x+dx$. This probability is simply the number of particles

Fig. 5.2 Concentration profiles of Brownian particles which were located on a thin slab at $t = 0$. The bell-shaped curve in a box represents the relative density of the particles for each point in the x-direction. The curve at the bottom illustrates the case in which the root-mean-square displacement at $t=3$ s equals $\sqrt{3}$ cm. Figure is adapted from B.H. Lavenda, *Brownian Motion*, Scientific American **252** (1985), pp. 56–67

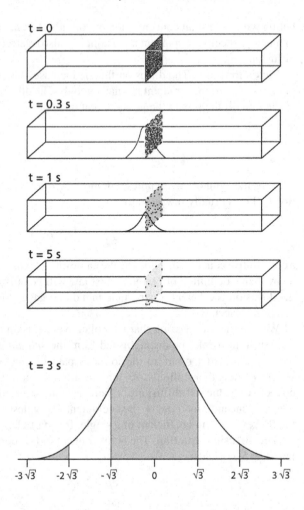

in that region, divided by the total number of particles in the original source. For a plate source (Fig. 5.2) with a an initial surface number density Γ_0 at $t = 0$ we have (see also Exercise 5.1)

$$P(x, t)\mathrm{d}x = \frac{\rho(x, t)}{\Gamma_0}\mathrm{d}x = \frac{\exp[-x^2/4Dt]}{2(\pi Dt)^{1/2}}\mathrm{d}x \qquad (5.15)$$

So the diffusion coefficient is all we need to predict where particles are located at any moment in time.

Diffusion plus convection. Often the randomizing diffusion or Brownian motion is accompanied by a *convection, i.e.* directed transport of particles by external forces such as gravity, an electric field or a flow field. For a concentration ρ of colloids each moving with a velocity $\vec{u}$, the convective particle flux is:

$$\vec{j}_c = \rho \vec{u} \tag{5.16}$$

Supplementing this convective flux with a diffusive flux $\vec{j}_d$ the continuity Eq. (5.7) for the colloid concentration becomes:

$$\frac{\partial \rho}{\partial t} = -\vec{\nabla} \cdot (\vec{j}_d + \vec{j}_c) = D\nabla^2\rho - \vec{\nabla} \cdot (\rho\vec{u}) \tag{5.17}$$

This is the convection–diffusion equation which will be employed in Sect. 6.2 to derive the Stokes-Einstein diffusion coefficient, and in Chap. 10 to analyze Brownian motion in an external force field.

Suppose a gravitational force propels particles with mass m at a speed $u = mg/f$ in the x-direction, the convective flux equals $j_c = \rho mg/f$. Then according to (5.17) the particle concentration in the x-direction changes in time as

$$\frac{\partial \rho}{\partial t} = D\nabla^2\rho - \frac{mg}{f}\nabla\rho = D\frac{\partial^2\rho}{\partial x^2} - \frac{mg}{f}\frac{\partial\rho}{\partial x} \tag{5.18}$$

Here g is the gravitational acceleration and f is the Stokes friction factor of a particle with mass m. Equation (5.18) will re-appear in the derivation of the diffusion coefficient in Chap. 6.

5.3 Stationary Diffusion

Solving the diffusion Eq. (5.14) is even for simple geometries mathematically involved. A considerable simplification occurs when the diffusion process enters a stationary (or steady) state in which the concentration profile of diffusing particles does not change in time. Then $\partial\rho/\partial t = 0$ so the Laplacian of the concentration in Eq. (5.14) is zero:

$$\nabla^2\rho = 0 \tag{5.19}$$

This result is also known as the Laplace equation. To solve it, the geometry and the boundary conditions of the diffusion problem must first be specified. For one-dimensional stationary diffusion in the x-direction we have the Laplace equation in the form:

$$\frac{d^2\rho}{dx^2} = 0, \tag{5.20}$$

Fick's first law (5.12) for the diffusion flux in one dimension reads:

$$j_d = -D\frac{d\rho}{dx} \tag{5.21}$$

Fig. 5.3 Sketch
accompanying the estimate
in Sect. 5.4 of the diffusion
coefficient for particles in a
dilute gas phase. Particles
diffuse from higher to lower
concentration across a
distance approximately equal
to the mean free path

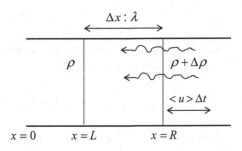

Note the minus sign: it is there because particles diffuse from high to low concentration so diffusion fluxes always go downhill a concentration profile. From (5.21) and (5.20) it follows that $dj_d/dx = 0$; an instance of the steady state condition (5.10).

Brownian motion in a tube. One simple geometry is a vessel with colloid concentration C_A connected by a narrow tube to another vessel with concentration $C_B < C_A$. Provided both vessels are large enough, Brownian motion of colloids in the narrow tube will not significantly change the colloid concentrations C_A and C_B and, consequently, a stationary flux of colloids will set up. From (5.21) and (5.20) we can infer that the stationary flux j_D is indeed a constant, with a magnitude that follows from integration of (5.21):

$$j_d = \frac{D(C_A - C_B)}{L} \tag{5.22}$$

Here L is the tube length, running from $x=0$ to $x=L$ where colloids enter or leave the vessel with concentration C_B. Measurement of a stationary diffusion flux for known colloid concentrations C_A and C_B allows the determination of the colloid diffusion coefficient D.

5.4   Diffusion in a Dilute Gas

In Sect. 6.1 Fick's first law (5.21) will be utilized by Einstein to find the diffusion coefficient of a solute particle in solution. Here we will employ Fick's law to estimate the diffusion coefficient for particles in a low-density gas phase. The estimate is based on the mean-free-path λ from Sect. 3.2, which is here assumed to be the distance that separates a particle number density $\rho+\Delta\rho$ at location $x=R$ from a lower concentration ρ at location $x=L$, see Fig. 5.3. If $<u>$ is the average particle speed, then the number of particles that travel in Δt seconds through area A from R to L is:

$$\frac{1}{2} \times \; < u > \; \Delta t \times A \times (\rho + \Delta\rho), \tag{5.23}$$

where a factor ½ is included because only half of the particles move in the *R-L* direction. The number of particles that moves in Δt seconds from *L* to *R* is:

$$\frac{1}{2} \times <u> \Delta t \times A \times \rho \qquad (5.24)$$

Consequently, the net transport per second through a unit area is

$$j_D = \frac{1}{2} <u> \times \Delta \rho \approx -\frac{1}{2} <u> \times \lambda \frac{\Delta \rho}{\Delta x}, \qquad (5.25)$$

taking into account that particle transfer occurs over a distance about equal to the mean free path. By comparison with Fick's law (5.21) we arrive at the following estimate for the diffusion coefficient:

$$D \approx \frac{1}{2} <u> \lambda \qquad (5.26)$$

The simple model depicted in Fig. 5.3 provides the correct scaling $D \sim <u> \lambda$ but the numerical factor is inaccurate; a much more involved derivation[1] yields:

$$D = \frac{1}{3} <u> \lambda = \frac{1}{3} \lambda^2 z \qquad (5.27)$$

Here z is the collision frequency on one particle, see Sect 3.2. From Chap. 3 we recall that the mean free path is also given by:

$$\lambda = \frac{1}{\rho \pi d^2} = \frac{kT}{p \pi d^2} \qquad (5.28)$$

Thus the diffusion coefficient of small molecules exceeds that of large molecules, and it decreases upon increase of the pressure or particle number density.

Gas viscosity. Diffusive particle transport quantified by diffusion coefficient *D* has associated with it diffusive transport of momentum, measured by the viscosity η. For a gas with mas density δ:

$$D = v = \frac{\eta}{\delta} \qquad (5.29)$$

Here v is the kinematic viscosity, see also the paragraph on momentum diffusion in Sect. 7.2. Hence the viscosity of a gas equals:

$$\eta = \frac{1}{3} <u> \lambda \delta \qquad (5.30)$$

---

[1] See f.e. J. Jeans, *An Introduction to the Kinetic Theory of Gases* (Cambridge University Press, 1962)

A gas of particles with mass m has a mass density $\delta = \rho m$ such that:

$$\eta = \frac{1}{3} < u > \frac{m}{\pi d^2}, \tag{5.31}$$

showing that, in contrast to the diffusion coefficient, gas viscosity is *independent* of the concentration and pressure of particles. One could have expected that a particle, when it has to plough through a particle swarm of increasing density, will experience a higher friction and viscosity. Maxwell confirmed his own prediction of the pressure independence by measuring gas viscosities at various pressures. The independence arises from two compensating effects: when gas density increases more molecules are available for momentum transport but they carry momentum less far because of the decrease of the mean free path.

Exercises

5.1 An infinite plate contains at time $t = 0$ a surface number density Γ_0 of Brownian particles. Diffusion produces a bell-shaped concentration profile that gradually flattens in time (Fig. 5.2). Verify that the profile

$$\frac{\rho(x, t)}{\Gamma_0} = \frac{\exp[-x^2/4Dt]}{2(\pi Dt)^{1/2}} \tag{5.32}$$

is a solution of Eq. (5.14) for diffusion in the x-direction.

5.2 Suppose at $t=3$ s, $\sqrt{< x^2 >} = \sqrt{3}$ cm. Calculate the probability that at $t = 3$ s a particle is found within $\sqrt{3}$ cm of the thin slab (Fig. 5.2) at $x = 0$.

5.3 Show from the profile in Exercise 5.1 that the number of diffusing particles in the box in Fig. 5.2 is conserved.

References

A very pedagogical textbook on the subject matter of this chapter is: R. F. Probstein, *Physicochemical Hydrodynamics; an Introduction* (2nd Edition, Wiley Europe, 1994).

For further informal discussion of surface integrals and the divergence theorem see: H. M. Schey, *Div, Grad, Curl and All That* (Norton, New York 1973).

Solutions for the diffusion equation (5.14), f.e. for the flat plate in Exercise 5.2, can be found in: J. Crank, *The Mathematics of Diffusion* (Oxford, University Press, 1964).

Chapter 6
Brownian Displacements

The trajectory of a Brownian particle is an erratic curve with the characteristic feature that the observed *distance* in a given time interval Δt depends on the magnification of the microscope (Fig. 6.1). Thus one cannot differentiate this distance unambiguously with respect to time to obtain a velocity. Instead we have to focus on the *displacement* of the particle, defined as the shortest distance between two positions of the colloid. How the squared displacement by diffusion grows in time was first figured out by Einstein.

6.1 Einstein for Chemists

> Prof. R. Lorentz has called to my attention, in a verbal communication, that an elementary theory of the Brownian motion would be welcomed by a number of chemists. Acting on this invitation, I present in the following a simple theory of the phenomenon.

Thus opened Albert Einstein (Fig. 6.2) his last publication[1] on the subject of Brownian motion. Writing for an audience of chemists, as we shall see below, apparently implied leaving out mathematics as much as possible. Below we will paraphrase Einstein's ingenuous argumentation, occasionally giving the floor to the man himself.

'Diffusion and Osmotic Pressure'[2]. Einstein draws a cylindrical vessel Z (Fig. 6.3) filled with a dilute solution, and divided by a movable semi-permeable piston M in regions A and B. If the solute concentration in A exceeds the concentration in B, an external force, directed to the left, is needed to maintain the piston in equilibrium. This force equals the difference between the two osmotic pressures

---

[1] A. Einstein, *Elementare Theorie der Brownschen Bewegung*, Z. für Electrochemie **14** (1908), 253–239. Quotations in this Chapter are English translations in A. Einstein, *Investigations on the Theory of the Brownian Motion*, (Ed. by R. Fürth, Dover 1956).

[2] Subtitles are from Einstein, in the paper in note 1.

© Springer Nature Switzerland AG 2018
A. P. Philipse, *Brownian Motion*, Undergraduate Lecture Notes in Physics,
https://doi.org/10.1007/978-3-319-98053-9_6

Fig. 6.1 Brownian motion
observed by Perrin for
mastic spheres (radius
0.53 μm) in water. Particle
positions were marked every
30 s. The side of a square in
A is about 3 micron. *Source*
J. Perrin, *Atoms* (transl. D. L.
Hammick), Constable &
Company Ltd, London, 1916

(a)

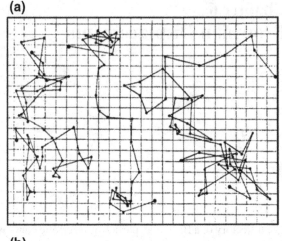

(b)

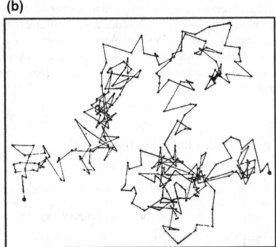

exerted by solutes on the piston. In absence of this external force, the piston will
move to the right until concentrations in *A* and *B* are equal. From this consideration
it follows, according to Einstein (see Footnote 1) that

> it is the forces of osmotic pressure that bring about the equalization of the concentrations in
> diffusion[3]; for we can prevent diffusion [...][4] by balancing the osmotic differences, which
> correspond to the differences of concentrations, by external forces acting on semi-permeable
> partitions.

Next Einstein imagines diffusion of dissolved substances taking place in Fig. 6.3
across the partition in cylinder *Z*. First he evaluates the 'osmotic forces' giving rise

---

[3]Here Einstein refers to collective or gradient diffusion that equalizes concentration differences.

[4]Here and later [...] indicates that in a quotation text has been skipped.

Fig. 6.2 Albert Einstein in ballistic motion

to this diffusion in a thin slice between planes E and E', see Fig. 6.3. Pressure π and π^* act on, respectively, planes E and E' so the net pressure on slice with thickness dx is

$$\pi - \pi^* \qquad (6.1)$$

From this net osmotic pressure it follows that

$$K = \frac{\pi - \pi^*}{dx} = -\frac{\pi^* - \pi}{dx} = -\frac{d\pi}{dx}, \qquad (6.2)$$

is the osmotic pressure gradient, corresponding to a force K acting on the solutes in a unit volume. Since the osmotic pressure is given by $\pi = \rho kT$, where ρ is the solute particle number density, it follows that the net osmotic force per particle equals

$$K = -\frac{kT}{\rho}\frac{d\rho}{dx} \qquad (6.3)$$

Einstein notes that "to calculate the motions due to diffusion to which these active forces can give rise, we must know how great a resistance the solvent offers to a

Fig. 6.3 Two drawings from
Einstein in his paper
*Elementare Theorie der
Brownschen Bewegung,*
Zeitschrift für Electrochemie
14 (1908), 253–239,
employed to derive the
diffusion coefficient, see
Sect. 6.1. Einstein uses for
osmotic pressure the symbol
p; in the text osmotic
pressure is denoted by *π*

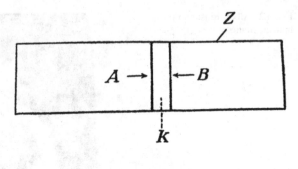

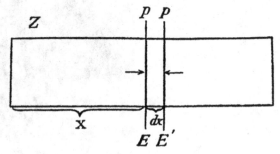

movement of the dissolved substance". This resistance is known because when a
force K acts on a particle, the particle's stationary speed u is given by:

$$u = \frac{K}{f},\qquad(6.4)$$

where f is the Stokes friction factor. The concentration difference in Z (Fig. 6.3)
creates a flux of particles, carried by diffusion through the thin slice between planes
E and E'. The magnitude of this particle flux follows from (6.4) and (6.3) as

$$\rho u = \rho \frac{K}{f} = - \frac{kT}{f}\frac{d\rho}{dx}\qquad(6.5)$$

Einstein remarks that the product ρu represents the amount of dissolved substance
"carried per second by diffusion through unit area of cross section", and then delivers
the following punch line:

"The multiplier of $d\rho/dx$ is therefore nothing else[5] but the coefficient of diffusion
D of the solution in question. We have, therefore, in general:

$$D = \frac{kT}{f},\qquad(6.6)$$

---

[5]The diffusion coefficient is defined as the ratio of flux to gradient, see also Sect. 5.2.

and, in the case when the diffusing molecules can be looked upon as spherical, and
large compared to the molecules of the solvent [...]:

$$D = \frac{kT}{6\pi \eta R} \tag{6.7}$$

In the last case, therefore, the coefficient of diffusion depends upon no other
constants characteristic of the substance in question but the viscosity η of the solvent
and the radius R of the molecule".

In the derivation of (6.6) Einstein makes no assumptions about the shape, size
or composition of the Brownian particles. The diffusion coefficient $D = kT/f$ applies,
for example, equally well to inorganic colloids of any shape, as well as to polymers
and proteins: details of shape and size only affect the friction factor. The latter can
only be calculated for a limited number of shapes—we will do the calculation for a
sphere in Chap. 8.

'**Diffusion and Irregular Motion of the Molecules**'. In the above line of rea-
soning for the diffusion coefficient D, Einstein does not commit himself to any
microscopic picture of particle diffusion. He merely evaluates the particle flux due
to a concentration gradient[6], after which D follows from Fick's first law.

His second derivation 'for chemists' concerns the displacement by diffusion and
here Einstein starts with a microscopic depiction of diffusion as a process in which
"single molecules of a liquid will alter their positions in the most irregular manner
thinkable". Then the argumentation proceeds as follows. Suppose particles diffuse
in the direction of the x-as of cylinder Z, in a time interval τ so short that solute
concentrations hardly change. Let Δ be the typical displacement of a particle in the
x-direction in time τ, a displacement that is just as frequently positive as negative.

Fig. 6.4 Einstein made use
of this drawing in the
derivation of the quadratic
displacement of Brownian
particles, as explained in
Sect. 6.1

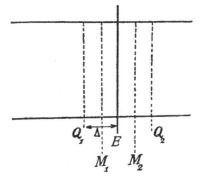

From left to right across plane E (Fig. 6.4) only solutes will diffuse that are located within a distance Δ at the left from E. Since only half of the solutes between planes Q and E makes a displacement $+\Delta$, the number of those particles is

$$\frac{1}{2}\rho_L\Delta, \tag{6.8}$$

where ρ_L is the mean solute number density in volume QE, *i.e.* the density in the middle layer M_1. Since the cross-section is unity, (6.8) is the number of solute particles in QE. The number of particles that diffuse through E from left to right in time τ equals

$$\frac{1}{2}\rho_R\Delta \tag{6.9}$$

Here ρ_R is the solute number density in middle layer M_2. The net number of particles diffusing from left to right in τ seconds is:

$$\frac{1}{2}\Delta(\rho_L - \rho_R) \tag{6.10}$$

The derivative of ρ with respect to x (the distance from the left cylinder-end) is:

$$\frac{d\rho}{dx} = \frac{\rho_R - \rho_L}{\Delta}, \tag{6.11}$$

so the number of particles from (6.10) can also be expressed as

$$-\frac{1}{2}\Delta^2\frac{d\rho}{dx} \tag{6.12}$$

Consequently, the number of particles diffusing per second across E is given by:

$$-\frac{1}{2}\frac{\Delta^2}{t}\frac{d\rho}{dx} \tag{6.13}$$

Just as in Eq. (6.5) we have here a particle flux due to a concentration gradient and also here the factor multiplying the gradient must be the diffusion coefficient:

$$D = \frac{1}{2}\frac{\Delta^2}{t} \tag{6.14}$$

Thus the diffusive displacement in time t is

$$\Delta = \sqrt{2Dt}, \qquad (6.15)$$

showing the characteristic square-root time dependence of the displacement by Brownian motion. Einstein remarks at this point that Δ actually should be replaced by the root-mean-squared displacement $<\Delta^2>^{1/2}$—as we will confirm in Sect. 6.2.

Remark on ion diffusion. On various occasions Einstein emphasizes that the solute particles are large compared to solvent molecules which justifies the use of the friction factor f derived for a solvent that behaves as a structure-less hydrodynamic continuum. For colloids on the diffusive time scale this continuum assumption is certainly justified. One would expect that for diffusion of molecular species, comparable in magnitude to water molecules, the continuum assumption breaks down. Diffusion of ions, nevertheless, obeys the Stokes-Einstein diffusion coefficient fairly well; for ions in water and water molecules themselves, typically[7] $D \sim 10^{-5}$ cm^2/s which is also the order of magnitude expected from (6.7) for the molecular M-sphere from Table 1.1, see also Exercise 1.1.

6.2   Translational Diffusion Coefficient from Equilibrium

The treatment 'for chemists' in the previous section is Einstein's informal account of his own first, more technical treatment of Brownian motion that was published in a physics journal[8]. In both treatments the ingredients of the diffusion coefficient are the thermal energy kT from Van't Hoff's law for solute particles, and the friction factor f for these same particles drifting in a viscous fluid. In addition, in both treatments the law for the mean-square-displacement (MSD) is found by focusing on the erratic motions of solute particles. In his 1905 paper Einstein found the MSD via Fick' second diffusion law, to which we return in Sect. 6.3. Here we first address Einstein's approach to calculate D by deriving two different expressions for the very same equilibrium density profile.

---

[7]A notable exception is the proton which diffuses much faster.

[8]A. Einstein, *Über die von der molekularkinetischen Theorie der Wärme geforderte Bewegung von in ruhenden Flüssigkeiten suspendierten Teilchen*, Ann. D. Physik **17**, (1905), pp. 549–560.

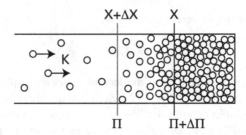

Fig. 6.5 Thermodynamic equilibrium between osmotic pressure π and external force K implies that net particle fluxes and net momentum flux are both zero, which entails the Stoke-Einstein diffusion coefficient $D = kT/f$, see Sect. 6.2

Suppose colloids in a dispersion experience an external force K, for example gravity or an electrical or magnetic force. The colloids accumulate in one part of the vessel (Fig. 6.5), which is counteracted by Brownian motion that tends to homogenize the particle distribution. In equilibrium the two tendencies balance, leading to a concentration profile $\rho(x)$ which remains constant in time. This thermodynamic equilibrium profile implies that in any volume element the net flux of both momentum and colloids is zero.

Momentum flux. The momentum flux results from two forces. The gradient in concentration produces a gradient in osmotic pressure π which corresponds to a force per unit volume of dispersion, also referred to as a *force density*. This osmotic force density is balanced by the external force which, per unit volume, equals K times the number density ρ of colloids. In equilibrium the sum of force densities (the net momentum flux) is zero:

$$\rho \vec{K} + \vec{\nabla} \pi = 0 \tag{6.16}$$

We now assume that the colloids do not interact such that they obey Van't Hoff's law $\pi = \rho kT$. Then for a one-dimensional profile that changes only in the x-direction, (6.16) simplifies to the differential equation:

$$\rho K + kT \frac{d\rho}{dx} = 0, \tag{6.17}$$

with the solution:

$$\rho(x) = \rho_0 \exp[-Kx/kT] \tag{6.18}$$

Here ρ_0 is the number density at $x = 0$. This is the Boltzmann distribution of non-interacting particles in an external force field. For colloids sedimenting in gravity, K can be identified as the weight of a colloid corrected for buoyancy. For that case Eq. (6.18) is also called the sedimentation-diffusion equilibrium profile.

Particle flux. We can describe equilibrium also in terms of the particle flux $\vec{j}$ which appears in the continuity equation

$$\frac{\partial \rho}{\partial t} = -\vec{\nabla} \cdot \vec{j} \tag{6.19}$$

Suppose the external force K propels particles at a stationary, constant speed u. The corresponding convective flux ρu has to be substracted from the diffusive flux due to Brownian motion, so the net flux in the x-direction is:

$$j_x = -D\frac{\partial \rho}{\partial x} - \rho u \tag{6.20}$$

Hence for convection and diffusion in the x-direction Eq. (6.19) becomes:

$$\frac{\partial \rho}{\partial t} = -\frac{\partial}{\partial x}\left[\rho u + D\frac{\partial \rho}{\partial x}\right] \tag{6.21}$$

This is the diffusion-convection equation for the time and spatial dependence of the colloid number density that we already encountered in Sect. 5.2. In a stationary state the concentration profile $\rho(x)$ is independent of time t:

$$0 = -\frac{d}{dx}\left[\rho u + D\frac{d\rho}{dx}\right], \tag{6.22}$$

implying that the total flux is a constant, independent of x. Since in thermodynamic equilibrium there can be no net particle flux, this constant must be zero:

$$\rho u + D\frac{d\rho}{dx} = 0, \tag{6.23}$$

which can be integrated to yield:

$$\rho(x) = \rho_0 \exp[-ux/D] \tag{6.24}$$

This equilibrium profile should be the same profile as in Eq. (6.18) which entails that the diffusion coefficient D equals:

$$D = \frac{u}{K}kT = \mu kT \tag{6.25}$$

The mobility coefficient μ in (6.25) is defined as the stationary speed per unit of applied force:

$$u = \mu K \tag{6.26}$$

Colloids with a stationary speed u in a viscous fluid experience a frictional force fu; thus $K=fu$ and the mobility μ is the inverse Stokes friction factor such that:

$$D = \frac{kT}{f} \tag{6.27}$$

6.3 Quadratic Displacements via Einstein's Diffusion Approach

The mean square displacement (MSD) of a Brownian particle can be found as follows via the diffusion equation[9]. Consider a particle which diffuses for a time t to reach a (positive or negative) displacement x with respect to the particle position at $t = 0$. We assume that there is no external force on the colloid, so positive and negative displacements occur with equal probability. The average for a large number of particles, also referred to as the *ensemble average*, of the displacement is therefore:

$$<x> = \int_{-\infty}^{+\infty} P(x, t)x \, dx = 0, \tag{6.28}$$

where $P(x, t)dx$ is the probability that after t seconds, a particle displacement is in the interval between x and $x + dx$. The function $P(x, t)$ is a probability *density* (with unit 1/m) normalized via

$$\int_{-\infty}^{+\infty} P(x,t) \, dx = 1, \tag{6.29}$$

which expresses that the probability to find a particle *somewhere* equals one. The average of the quadratic displacement is calculated as follows. The probability to find a particle at a certain location x is proportional to the particle concentration $\rho(x, t)$ at that location:

$$P(x, t) \propto \rho(x, t) \tag{6.30}$$

---

[9]The derivation is in essence that of Einstein's 1905 paper; one difference is that we do not invoke the explicit solution of the diffusion equation, which simplifies the treatment.

This concentration is the solution of the diffusion equation from Chap. 5, which reads for diffusion in the x-direction:

$$\frac{\partial}{\partial t}\rho(x,t) = D\frac{\partial^2}{\partial x^2}\rho(x,t) \tag{6.31}$$

where D is the diffusion coefficient. Substitution of (6.30) yields for the probability density:

$$\frac{\partial}{\partial t}P(x,t) = D\frac{\partial^2}{\partial x^2}P(x,t), \tag{6.32}$$

which allows us to evaluate the ensemble average of the quadratic displacement via the integration:

$$<x^2> = \int_{-\infty}^{+\infty} P(x,t)x^2 dx; \quad \text{for } t \gg \tau_{MR} \tag{6.33}$$

This mean-square-displacement (MSD), applies for colloids that have entered the diffusive time regime, $i.e.$ the time t is much larger than the momentum relaxation time τ_{MR} discussed in Chap. 4. We are interested in the change of the MSD with time on this diffusive time scale:

$$\frac{d}{dt}<x^2> = \int_{-\infty}^{+\infty} \frac{\partial}{\partial t}P(x,t)x^2 dx = D\int_{-\infty}^{+\infty}\left[\frac{\partial^2}{\partial x^2}P(x,t)\right]x^2 dx \tag{6.34}$$

For physical reasons we can expect the distribution function $P(x,t)$ to have the following properties. Since steps in $+x$ and $-x$ directions are equally probable the function will be symmetric such that $P(+x,t) = P(-x,t)$. In addition, moving away from the origin both $P(x,t)$ and its first and second derivative will all asymptote monotonically to zero in the limits $x \rightarrow \pm\infty$. In other words, $P(x,t)$ must have the bell-shape curve depicted in Fig. 5.2. Given these properties of $P(x,t)$, two integrations by parts of (6.34) yield (Exercise 6.2):

$$\frac{d}{dt}<x^2> = 2D \tag{6.35}$$

which results in the Einstein equation for the average quadratic displacement:

$$<x^2> = 2Dt; \quad \text{for } t \gg \tau_{MR} \tag{6.36}$$

Note that this result has been obtained without solving the diffusion equation to find an explicit expression for $P(x,t)$, see also Exercise 6.1. For a randomly diffusing colloid there is no distinction between x, y and z directions (the random-motion postulate from kinetic theory) so $< x^2 >=< y^2 >=< z^2 >$. Thus the average quadratic radial displacement for a colloid diffusing in any direction $\vec{r}$ from a central point is given by:

$$<r^2> = 6\,Dt; \quad r^2 = x^2 + y^2 + z^2 \tag{6.37}$$

At this point the idea could arise that we can define an effective speed u_{eff} of the colloid by differentiating the distance

$$\sqrt{<r^2>} = \sqrt{6Dt} \tag{6.38}$$

with respect to time to obtain:

$$u_{\text{eff}} = \frac{\mathrm{d}\sqrt{<r^2>}}{\mathrm{d}t} = \sqrt{\frac{3D}{2t}} \tag{6.39}$$

This effective speed, however, diverges when time t approaches zero. What is going wrong here is that the expression for the MSD in (6.37) is only valid on the diffusion time scale $t \gg \tau_{\text{MR}}$; for times comparable to the momentum relaxation time the MSD has a time dependence that differs from (6.37), as will be shown in the next Sect. 6.4 on the Langevin equation.

6.4   Brownian Motion from Newtonian Mechanics

Einstein's treatment of Brownian motion in Sect. 6.3 is an analysis from the view point of the time-dependent diffusion of colloids: an ensemble of Brownian parti-cles, initially concentrated in the origin, diffuses out as time proceeds, leading to concentration profiles that follow from Fick's second law in Eq. (6.31), with average mean-square-displacements given by (6.36). Since the diffusion approach considers times far beyond the colloid's momentum relaxation time, the microscopic details underlying Brownian displacements—fluctuating forces exerted on particles by ther-mal shocks from the surrounding fluid—remain concealed. Precisely these details form the bases of a treatment of Brownian motion due to Paul Langevin (1872–1946), which he published a few years[10] after Einstein's *annus mirabilis* 1905.

The Langevin equation. Langevin examined Brownian movement from the per-spective of the *forces* acting on a colloid. The total force on one colloid, Langevin argued, can be separated into two parts: one is an *average* force due to the surround-ing fluid and the other part is the *transient* force that a colloid experiences due to fluid pressure fluctuations in its immediate vicinity. The average force stems from the fluid's viscosity and equals the frictional force $f\,\mathrm{d}x/\mathrm{d}t$ exerted on a particle moving

---

[10]M. P. Langevin, *Sur la theorie du mouven Brownien*, C. R. Acad. Sci. (Paris) 146, 530–533 (1908).

in the x-direction at a speed dx/dt. According to Newton's second law the total force on a particle equals the mass of the particle times its acceleration:

$$m\frac{d^2x}{dt^2} = -f\frac{dx}{dt} + F(t); \quad \langle F(t)\rangle = 0 \tag{6.40}$$

Here $F(t)$ is the fluctuating component of the total force, a component that on average is zero. The equation of motion (6.40) of a colloid is also known as the *Langevin equation*. We are interested in squared displacements x^2 rather than x. To get x^2 into the Langevin equation we multiply (6.40) by x/m and employ the identities

$$x\frac{dx}{dt} = \frac{1}{2}\frac{dx^2}{dt}; \quad x\frac{d^2x}{dt^2} = \frac{1}{2}\frac{d^2x^2}{dt^2} - \left(\frac{dx}{dt}\right)^2, \tag{6.41}$$

to arrive at the Langevin equation in the form:

$$\frac{1}{2}\frac{d^2x^2}{dt^2} - \left(\frac{dx}{dt}\right)^2 = -\frac{1}{2\tau_{MR}}\frac{dx^2}{dt} + \frac{xF(t)}{m}; \quad \tau_{MR} = \frac{m}{f}, \tag{6.42}$$

where τ_{MR} is the momentum relaxation time that we already encountered in Chap. 4. Now Eq. (6.42) holds for *one* particular particle only; for many identical particles in the fluid we take, for a given moment in time, the ensemble average[11] of the terms in (6.42):

$$\frac{1}{2}\frac{d^2\langle x^2\rangle}{dt^2} - \langle(dx/dt)^2\rangle = -\frac{1}{2\tau_{MR}}\frac{d\langle x^2\rangle}{dt} + \frac{\langle xF(t)\rangle}{m} \tag{6.43}$$

Here $\langle x^2\rangle$ is the mean-square-displacement, hereafter abbreviated as MSD. When the slow, heavy colloid has moved a distance x, the force $F(t)$ has already randomly fluctuated many times. Thus we can assume that x and $F(t)$ are uncorrelated on the time scale of colloid motion, entailing that the average of their product is zero:

$$\langle xF(t)\rangle = \langle x\rangle\langle F(t)\rangle = 0 \tag{6.44}$$

In addition, the mean-square speed of the colloids in the x-direction follows from equipartition as

$$\langle(dx/dt)^2\rangle = \langle v_x^2\rangle = \frac{kT}{m} \tag{6.45}$$

---

[11] A sequence of observations in time on a *single* particle will yield the same averages.

On substitution of (6.44) and (6.45) the Langevin Eq. (6.42) transforms to the following differential equation for the MSD:

$$\frac{d^2\langle x^2\rangle}{dt^2} + \frac{1}{\tau_{MR}}\frac{d\langle x^2\rangle}{dt} = \frac{2kT}{m} \tag{6.46}$$

Integration yields for the time derivative of the MSD:

$$\frac{d\langle x^2\rangle}{dt} = \frac{2kT}{f}(1 - \exp[-t/\tau_{MR}]) \tag{6.47}$$

This solution satisfies the boundary condition that the initial rate of displacement of the colloid is zero: $d < x^2 > /dt = 0$ at $t = 0$.

Mean square displacement. Upon integration of (6.47) we obtain for the MSD:

$$\langle x^2\rangle = \frac{2kT}{f}[t - \tau_m(1 - \exp[-t/\tau_{MR}])] \tag{6.48}$$

In the time dependence of the MSD we can discern two regimes. At times $t \ll \tau_{MR}$ Eq. (6.48) reduces to[12]

$$\langle x^2\rangle = \frac{kT}{m}t^2 = <v_x^2>t^2; \quad 0 < t \ll \tau_{MR} \tag{6.49}$$

This equation describes a type of motion that is consistent with a particle with constant speed u that covers in uniform, linear flight a distance s in time t:

$$s = u\,t \tag{6.50}$$

Such a free flight is also referred to as ballistic motion: an inertial motion that is governed by particle mass only. The particle's inertia is taken into account by the exponential term in (6.48) which vanishes at times much larger than the momentum relaxation time:

$$\langle x^2\rangle = 2\frac{kT}{f}t = 2Dt; \quad t \gg \tau_{MR} \tag{6.51}$$

This result for the time dependence of the MSD is the same as the one found via Einstein's diffusion approach in Sect. 6.3. However, via the Langevin equation we also recover here the diffusion coefficient $D=kT/f$.

Mass memory loss. The physical picture emerging from the Langevin equation is as follows. A colloid, when we start to clock it at $t=0$, begins its journey as a ballistic object governed by the inertial term md^2x/dt^2 in its equation of motion (6.40).

---

[12]Make use of the Taylor expansion $\tau_m(1 - \exp[-t/\tau_m])] = t + (1/2)t^2/\tau_m + \dots$

However, in the process of making many momentum relaxation steps the particle gradually loses 'memory' of its mass eventually leading to Brownian diffusion, with the MSD from (6.51) that has become independent of particle mass. In the diffusive regime the net force on the colloid is zero, so the erratic deflections in the trajectory (Fig. 6.1) of the 'force-free' particles are not velocity changes but random thermal steps.

Thus from observations of Brownian movements in a fluid under a microscope—with observation times $t \gg \tau_{MR}$—we can only deduce the size of the particles (via the diffusion coefficient) but not the particle mass. The latter requires measurement of the microscopic colloid speed in (6.49) which cannot be done with microscopy on a dispersion of colloids in a liquid phase. However, for colloids in a dilute *gas* phase microscopic speeds can, in principle, be measured, as mentioned at the end of Chap. 4.

6.5 Angular Displacements from a Diffusion Equation

The MSAD. Consider a label diffusing on the surface of a unit sphere with radius $R = 1$ (Fig. 6.6). The unit vector $\hat{u}(t)$ denotes the position of the label or, equivalently, the orientation of an axis through the label and the sphere center. The angular displacement of the label at time t is defined by the vector $\hat{u}(t) - \hat{u}_0$, where $\hat{u}_0$ marks the label position at $t = 0$. The mean of the squared angular displacement, hereafter referred to as the MSAD, is given by:

$$< \left(\hat{u}(t) - \hat{u}_0 \right)^2 > = < \hat{u}_0 . \hat{u}_0 > + < \hat{u}(t) . \hat{u}(t) > -2 < \hat{u}(t) . \hat{u}_0 > \quad (6.52)$$

Since the dot product of a unit vector with itself equals unity we find for the MSAD;

$$< \left(\hat{u}(t) - \hat{u}_0 \right)^2 > = 2 - 2 < \hat{u}(t) . \hat{u}_0 > = 2 - 2 < \cos \theta > ; \quad t \gg \tau_{AR} \quad (6.53)$$

Here θ is the angle between the unit vectors $\hat{u}(t)$ and $\hat{u}_0$. Note that (6.53) is only valid on the diffusional time scale, *i.e.* at times much larger than the time τ_{AR} for the angular momentum to relax, see Chap. 4. At $t=0$ all labels have the same orientation $\hat{u}_0$ such that $< \cos \theta > = 1$ and the MSAD is zero. By rotational Brownian motion the orientations gradually randomize until $< \cos \theta > = 0$ (Exercise 6.6) signifying that all orientations occur with equal probability such that

$$< \left(\hat{u}(t) - \hat{u}_0 \right)^2 > = 2; \quad t \to \infty \quad (6.54)$$

So the MSAD reaches a plateau equal to the sum of the two unit vector dot products in (6.52). Such a plateau is in marked contrast to the MSD for translational Brownian motion that increases without limit as time goes by. The plateau signifies that when

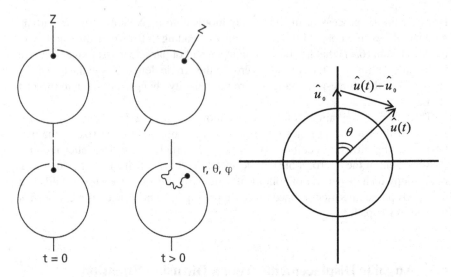

Fig. 6.6 Due to Brownian motion the direction of the z-axis (fixed to the spheres) fluctuates in time, which may be represented by a label that diffuses on the sphere surface from its initial position on the z-axis to a position (r, θ, ϕ) at time t. The displacement of the label is marked by the vector $\hat{u}(t) - \hat{u}_0$, where $\hat{u}_0$ is the label's position at $t = 0$

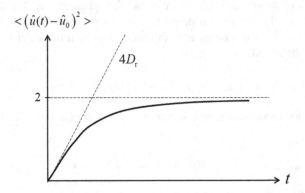

Fig. 6.7 The mean square angular displacement (MSAD) for diffusing labels on a sphere, see Fig. 6.5. At short times $D_r t \ll 1$ the MSAD grows linearly in time, just as for labels in a 2-d plane. Whereas at larger times a Brownian particle on an true 2-d plane would wander off to infinity the MSAD of the label on a sphere surfaces approaches a constant, which is further discussed in Sect. 6.5

we observe the Brownian movements of a label on a sphere (Fig. 6.7) we cannot tell how often the label already has probed the sphere surface. In other words, the label just turns in a (very erratic) circle whereas Brownian particles in a large bulk wander off to infinity.

Orientational decay. To assess how the MSAD relaxes in time to the plateau in (6.54), we have to find the time dependence of $< \cos \theta >$, for which we need the orientational distribution function $P(\theta, t)$. In complete analogy with Eq. (6.32) for $P(x,t)$ we have:

$$\frac{\partial}{\partial t} P(\theta, t) = D_r \nabla^2 P(\theta, t) \tag{6.55}$$

Here D_r is the rotational diffusion coefficient, actually a frequency with dimension 1/s. Note that $P(\theta, t)$, a probability density with dimension m^{-2}, does not depend on the angle ϕ (see also Fig. 6.6) so for a unit sphere with radius $R = 1$ the normalization condition is:

$$2\pi \int_{\theta=0}^{\pi} P(\theta, t) \sin\theta d\theta = 1 \tag{6.56}$$

The average of $\cos\theta$ follows from:

$$< \cos\theta > = 2\pi \int_{\theta=0}^{\pi} P \sin\theta \cos\theta d\theta; \quad P = P(\theta, t) \tag{6.57}$$

For the time-derivative we can write, making use of the diffusion Eq. (6.55)

$$\frac{d}{dt} < \cos\theta > = 2\pi \int_{\theta=0}^{\pi} \frac{\partial P}{\partial t} \sin\theta \cos\theta \, d\theta$$

$$= 2\pi D_r \int_{\theta=0}^{\pi} \frac{1}{\sin\theta} \left[\frac{\partial}{\partial\theta} \left(\sin\theta \frac{\partial P}{\partial\theta} \right) \right] \sin\theta \cos\theta \, d\theta \tag{6.58}$$

Note that we only need the θ-dependent part of the Laplace operator in spherical coordinates (see Appendix B). Partial integration of the integral in (6.58) yields:

$$\left[\cos\theta \, \sin\theta \frac{\partial P}{\partial\theta} \right]_{\theta=0}^{\pi} - \int_{\theta=0}^{\pi} \sin\theta \frac{\partial P}{\partial\theta} d(\cos\theta) \tag{6.59}$$

The bracket term equals zero and the remaining integral, after a second partial integration, turns out to equal:

$$-2 \int_{\theta=0}^{\pi} P \sin\theta \cos\theta d\theta = -\frac{1}{\pi} < \cos\theta > \tag{6.60}$$

The time-derivative in (6.58) therefore becomes

$$\frac{\mathrm{d}}{\mathrm{d}t} < \cos\theta > = -2D_r < \cos\theta > \qquad (6.61)$$

with the solution

$$< \cos\theta > = \exp[-t/\tau_{RR}]; \quad \tau_{RR} = \frac{1}{2D_r} \qquad (6.62)$$

Thus the orientation of initially aligned, non-interacting anisometric particles decays exponentially in time. The rotational relaxation time τ_{RR} is determined by the rotational diffusion coefficient of the particles in question (spheres, rods, platelets, magnetic dipoles etc.). Thus for the time dependence of the MSAD (6.53) we find:

$$< \left(\hat{u}(t) - \hat{u}_0\right)^2 > = 2(1 - \exp[-2D_r t]), \quad \text{for } t \gg \tau_{AR} \qquad (6.63)$$

Figure 6.7 depicts a sketch of the change of the MSAD in time. At short times such that $D_r t \ll 1$ Eq. (6.63) simplifies to:

$$< \left(\hat{u}(t) - \hat{u}_0\right)^2 > = < \theta^2 > = 4D_r t, \quad \text{for } D_r t << 1 \qquad (6.64)$$

This result is equivalent to translational diffusion on a two dimensional plane for which $< z^2 > = 4Dt$ where z is a two-dimensional displacement. Indeed, at short times the diffusive label has not probed the curvature of the sphere's surface yet. Gradually, however, the label discovers it is stumbling around on a sphere instead of on a flat plane and at longer times $D_r t \gg 1$, the MSAD levels to the plateau already identified in (6.54).

6.6   The Rotational Diffusion Coefficient

The diffusion coefficient $D = kT/f$ for translational Brownian motion was found from the distribution of particle *positions* resulting from the equilibrium between diffusion and an external *force*. The rotational diffusion coefficient D_r can be derived in a very similar manner from the distribution of particle *orientations* in response to an external *torque*. To this end we consider a collection of independent direction vectors as in Fig. 6.8, each representing an orientation angle θ with respect to the axis at $\theta = 0$. The vectors rotate with the same angular velocity:

$$\Omega = \frac{\mathrm{d}\theta}{\mathrm{d}t}[\text{rad s}^{-1}], \qquad (6.65)$$

towards $\theta = 0$. This rotation is caused by a torque T_0, the physical nature of which we do not have to specify. One can choose a rotating external magnetic field acting on magnetic dipoles, or a shear flow aligning rods but for the argumentation here

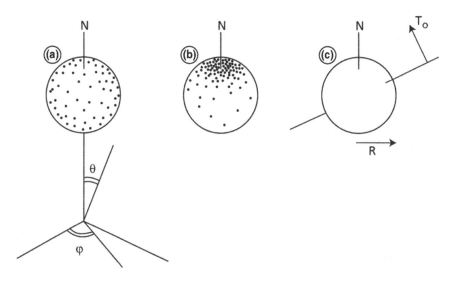

Fig. 6.8 Diffusing labels on a sphere surface (**a**) accumulate at the north pole (**b**) since all labels represents an axis subjected to the same constant torque T_0 as indicated in (**c**). Rotational diffusion tends to randomize axis orientations and counteracts the angular convection due to T_0. In equilibrium the distribution of labels is given by Eqs. (6.69) or (6.71)

this choice is irrelevant. All we ask from the torque is to sustain a constant angular velocity, which is given by:

$$\Omega = \mu_r T_0 \tag{6.66}$$

Here μ_r is the rotational mobility defined as the steady angular velocity per unit of applied torque; note the analogy between μ_r and the translational mobility μ in Eq. (6.25).

Angular flux. The number of vectors rotating per second, the convective angular flux, is given by:

$$j_\omega(\theta) = -\Omega\rho(\theta) = -\mu_r T_0 \rho(\theta), \tag{6.67}$$

in which $\rho(\theta)$ represents an orientation density, *i.e.* the number of direction vectors per unit angle. Due to this angular flux, vectors accumulate near $\theta = 0$, so a gradient $d\rho(\theta)/d\theta$ is formed which attempts to relax by rotational diffusion. The corresponding diffusive flux

$$j(\theta) = -D_r \frac{d\rho(\theta)}{d\theta}, \tag{6.68}$$

defines the rotational diffusion coefficient D_r of the independent, freely moving vectors. At equilibrium $j(\theta) + j_\omega(\theta) = 0$, and after integration we find the angular equilibrium profile:

$$\rho(\theta) = \rho(\theta = 0) \exp[-\mu_r T_0 \theta / D_r], \qquad (6.69)$$

which is the equivalent of (6.24) for the distribution of particle positions. Next we derive the angular equilibrium profile along a different route that employs work.

Angular work. The work done by the torque to achieve an angular displacement θ starting from $\theta = 0$ equals

$$w(\theta) = \int\limits_0^\theta T_0 d\theta' = T_0 \theta; \quad T_0 = \text{constant} \qquad (6.70)$$

The Boltzmann distribution of the orientations is therefore:

$$\rho(\theta) = \rho(\theta = 0) \exp[-w(\theta)/kT] = \rho(\theta = 0) \exp[-T_0 \theta / kT] \qquad (6.71)$$

The angular equilibrium distributions (6.69) and (6.71) are identical which entails that the rotational diffusion coefficient is given by:

$$D_r = \mu_r kT \qquad (6.72)$$

This result is still independent of the medium in which a particle is rotating. When the medium is a viscous fluid, the mobility is the inverse of the hydrodynamic Stokes friction factor f_r:

$$D_r = \frac{kT}{f_r} \qquad (6.73)$$

Take note of the close analogy between translational and rotational diffusion: thermal energy kT drives the random spreading of, respectively, positions and orientations, which is resisted by the same Stokes friction that also opposes, respectively, linear and angular drift in an external field.

Exercises

6.1   Substitute the solution of the diffusion equation from Exercise 5.1 in Chap. 5, in Eq. (6.33) to verify that the law for quadratic displacement in (6.36) is indeed correct.

6.2   Evaluate the integral in Eq. (6.34). Which assumptions do you have to make for $P(x,t)$? Why are they physically plausible?

6.3   Where in the derivation of the diffusion coefficient Eq. (6.25) it is assumed that the particles do not interact?

6.4   Estimate the time it would take oxygen molecules to diffuse in water ($D = 18 \times 10^{-6}$ cm^2/s) at room temperature a distance equal to (1) the typical thickness of a bacteria; (2) the typical thickness of a human being. Verify that diffusive transport of oxygen from the environment to the lungs is not an alternative to oxygen transport by red blood cells. Do you expect that an oxygen molecule in air diffuses much slower or much faster than in water? See for example S. Vogel *"Life's Devices; the physical world of animals and plants"* (Princeton University Press, 1988).

6.5   Discuss the validity of the rotational diffusion coefficient in Eq. (6.73) for non-spherical particles.

6.6   Show that, when orientations have randomized, $< \cos\theta > = 0$.

6.7   Colloids performing Brownian motion are force-free, see f.e. the discussion at the end of Sect. 6.4. In Sect. 6.1, however, Einstein evaluates the 'osmotic force' per colloid to derive the diffusion coefficient. Is there something going wrong in Einstein's argument? If not, why not?

6.8   Verify all steps in the derivation of the MSD (6.48) from the Langevin Eq. (6.40).

References

An English translation of Einstein's papers on Brownian motion can be found in: A. Einstein, *Investigations on the Theory of the Brownian Motion*, (Ed. by R. Fürth, Dover 1956).

Indispensable scientific biography of Albert Einstein, including an elucidating analysis of Einstein's papers on Brownian motion: A. Pais, *Subtle is the Lord* (Oxford University Press, 1982).

More extensive treatments of Brownian motion and other transport properties of colloids can be found in: W. B. Russel, D. A. Saville and W. R. Schowalter, *Colloidal Dispersions* (Cambridge, 1995), and J. K. G. Dhont, *An Introduction to Dynamics of Colloids* (Elsevier, Amsterdam, 1996).

For an English translation of Langevin's 1908 paper see: D. S. Lemons and A. Gythiel Am. J. Phys. **65** (11), November 1997.

Brownian motion in its wider context of stochastic processes is treated in: N. G. van Kampen, *Stochastic Processes in Physics and Chemistry*, (North Holland, 1981) and S. Chandrasekhar, *Stochastic Problems in Physics and Astronomy,* Rev. Mod. Phys. 15, 1 (1993).

Rotational diffusion is treated by Peter Debye in his classic *Polar Molecules* (Dover Publication, reprint of the 1929 Reinhold edition).

Chapter 7
Fluid Flow

Brownian motion in a liquid medium[1] is a dual process: the thermal motion of colloids in any direction is, owing to the liquid's incompressibility, accompanied by oppositely directed flow in the colloid's surroundings. In Chap. 4 we have seen that on times much beyond the time scale τ_C for molecular collisions, colloids experience a structure-less, continuous fluid characterized only by viscosity η and mass density δ. Thus we have to specify the mobility μ in the diffusion coefficient $D = \mu kT$ as a *hydrodynamic* mobility or its inverse, a hydrodynamic friction factor f. In this Chapter we will introduce hydrodynamics up to the level needed to calculate in Chap. 8 friction factors for translation and rotation of a solid sphere. Before eagerly plunging into the intricacies of hydrodynamics, we will first outline the challenging problem that a friction factor presents us.

Friction factors. When a liquid is stirred, fluid layers slide along each other—a process also referred to as *shear flow*—which causes dissipation of energy. This dissipation manifests itself as the viscosity of the liquid: the higher this viscosity, the more energy in the form of work has to be invested to maintain liquid motion. When an object is displaced in a fluid, one consequence is the occurrence of pressure forces because liquid has to be pushed out of the way. The other effect is induction of shear flows because liquid sticks to some extent to the object's surface giving rise to tangential shear forces. What we need to calculate is the total hydrodynamic force the fluid exercises on the moving object. This force equals the magnitude K of an external force that has to be exerted on the object to sustain its constant speed u. The friction factor f then follows as the ratio of the applied force and the steady speed $f = K/u$. This factor depends on the shape and size of the object in question, the simplest case being a smooth, undeformable solid sphere.

---

[1] To which there are notable exceptions such as colloids in a gas (aerosol) and the inorganic colloids that inhabit in vast numbers interstellar dust clouds.

© Springer Nature Switzerland AG 2018
A. P. Philipse, *Brownian Motion*, Undergraduate Lecture Notes in Physics,
https://doi.org/10.1007/978-3-319-98053-9_7

7.1 Fluid Velocity Fields

We begin with a general description of fluid flow. The flow velocity $\vec{u}$ in a fluid at position vector $\vec{r} = [x, y, z]$ and time t

$$\vec{u} = \vec{u}(\vec{r}, t), \tag{7.1}$$

has three Cartesian components u, v and w. Thus Eq. (7.1) is a shorthand for the vector function:

$$\vec{u} = [u(x, y, z, t), \; v(x, y, z, t), \; w(x, y, z, t)] \tag{7.2}$$

Finding the flow field $\vec{u}$ is the main task in a flow problem, because $\vec{u}$ tells us what all elements of the fluid are doing at any time t. A simplification is a *steady flow* in which both magnitude and direction of $\vec{u}$ are constant at any fixed point in space. Thus

$$\frac{\partial \vec{u}}{\partial t} = 0, \tag{7.3}$$

defines steady flow. Further, in many flow problems of interest (such as flow in a tube or past a sphere) $\vec{u}$ is independent of one or two spatial coordinates. For example, two-dimensional steady flow has the form:

$$\vec{u} = [u(x, y), v(x, y), 0] \tag{7.4}$$

With respect to the direction of the flow vector $\vec{u}$, an important concept is that of the *streamline* which is a curve that at any point has the same direction as $\vec{u}$. For a steady flow the streamline pattern does not change in time. Nevertheless, even though $\vec{u}$ is constant at a fixed point in space, the flow velocity may change for a particular fluid element traveling along its streamline.

The material derivative. We must clearly distinguish the fate of a blob of fluid which 'follows the flow' along a stream line, from what happens to fluid in a volume element that is fixed in space. This distinction also appears in the notation for derivatives of fluid properties. Suppose the function $f = f(x, y, z)$ denotes some property of the moving fluid such as its mass density or a component of velocity $\vec{u}$. According to the chain rule:

$$df = \frac{\partial f}{\partial t} dt + \frac{\partial f}{\partial x} dx + \frac{\partial f}{\partial y} dy + \frac{\partial f}{\partial z} dz \tag{7.5}$$

The total rate of change in f is therefore:

$$\frac{df}{dt} = \frac{\partial f}{\partial t} + \frac{\partial f}{\partial x}\frac{dx}{dt} + \frac{\partial f}{\partial y}\frac{dy}{dt} + \frac{\partial f}{\partial z}\frac{dz}{dt} \tag{7.6}$$

Suppose we measure f in a volume element rigidly attached to our boat, which is located at a position x, y, z. If this position is fixed then, according to Eq. (7.6), df/dt is the change in f in time. However, if we tour around in the fluid the rate of change in f also depends on the components dx/dt etc. of the boat velocity. Only if the boat (with its engine switched off) passively follows a streamline these components equal the components of the flow velocity $\vec{u}$:

$$\frac{dx}{dt} = u, \quad \frac{dy}{dt} = v, \quad \frac{dz}{dt} = w \tag{7.7}$$

For this particular case of 'following the fluid' the notation D/Dt is used instead of d/dt:

$$\frac{Df}{Dt} = \frac{\partial f}{\partial t} + u\frac{\partial f}{\partial x} + v\frac{\partial f}{\partial y} + w\frac{\partial f}{\partial z} = \frac{\partial f}{\partial t} + (\vec{u} \cdot \vec{\nabla})f \tag{7.8}$$

Here $(\vec{u} \cdot \vec{\nabla})$ is the vector product of the fluid velocity and the *differential operator* $\vec{\nabla}$ ('del'), defined in Cartesian coordinates as:

$$\vec{\nabla} = \sum_i \vec{\delta}_i \frac{\partial}{\partial x_i} \tag{7.9}$$

The derivative

$$\frac{D}{Dt} = \frac{\partial}{\partial t} + \vec{u} \cdot \vec{\nabla} \tag{7.10}$$

is known as the *material* or *substantial* derivative. Note that $\partial/\partial t$ denotes a change in time at a position that is fixed relative to the flow field, whereas the derivative $(\vec{u} \cdot \vec{\nabla})$ measures a change following a stream line.

Stream functions. From the meaning of the derivative $(\vec{u} \cdot \vec{\nabla})$ it follows that whenever

$$(\vec{u} \cdot \vec{\nabla})f = 0, \tag{7.11}$$

f is constant along a streamline; then f is a called *stream function*, because its value generates a streamline. Note that f might be a different constant on different streamlines, just as isobars in the weather forecast represent different but constant pressures. Stream functions will be needed later to analyze viscous flow past a translating sphere. By applying Eq. (7.8) to the components u, v and w of the fluid velocity the *acceleration* of a fluid element at position $\vec{r}$ is found to be:

$$\frac{D\vec{u}}{Dt} = \frac{\partial \vec{u}}{\partial t} + (\vec{u} \cdot \vec{\nabla})\vec{u} \tag{7.12}$$

Here $(\vec{u} \cdot \vec{\nabla})\vec{u}$ is a vector (see Appendix B) describing changes in $\vec{u}$ for a fluid blob traveling along a streamline. The acceleration in Eq. (7.12) is due to forces on the blob which will be identified below.

7.2   The Navier-Stokes Equation

Consider a surface S enclosing a region V in the fluid, with a normal unit vector $\vec{n}$ pointing outwards. The flow velocity component along the normal is $\vec{u}.\vec{n}$ so the net fluid volume leaving V in unit time is

$$\int_S \vec{u}.\vec{n}\,\mathrm{d}S \tag{7.13}$$

For an incompressible fluid this integral must be zero because there can be no net gain or loss of fluid volume. Using the divergence theorem (see Appendix B) we find

$$\int_S \vec{u}.\vec{n}\,\mathrm{d}S = \int_V \vec{\nabla}.\vec{u}\,\mathrm{d}V = 0, \tag{7.14}$$

which implies that

$$\vec{\nabla}.\vec{u} = 0 \tag{7.15}$$

anywhere in the fluid: the divergence of the flow is zero for any volume element in V. Note that (7.15) is the continuity equation for an incompressible fluid which we already encountered in Chap. 5. Next we consider the effect of the pressure $p = p(x, y, z)$ in the fluid. This pressure is a scalar function so the force K on a surface element $\mathrm{d}S$ is:

$$\vec{K} = -p\vec{n}\,\mathrm{d}S, \tag{7.16}$$

with a minus sign because $\vec{n}$ points out of the region V. The net force on the region is, using the divergence theorem (see Appendix B):

$$-\int_S p\vec{n}\,\mathrm{d}S = -\int_V \vec{\nabla}p\,\mathrm{d}V \tag{7.17}$$

If $\vec{\nabla}p$ is continuous it will be almost constant over a sufficiently small blob of volume ∂V. The net pressure force on the blob due to the surrounding fluid is therefore $-\vec{\nabla}p(\partial V)$. The gravitational force on the blob with mass density δ equals

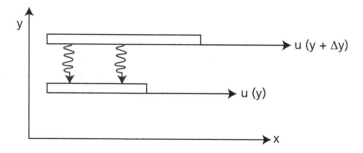

Fig. 7.1 Momentum transport in the y-direction, from a fast fluid layer at height $y + \Delta y$ to a slower fluid layer at y, gives rise to a shear force in the x-direction at height y

$\delta \vec{g}(\partial V)$. The sum of forces must equal the product of the blob's mass $\delta(\partial V)$ and its acceleration, so we obtain:

$$\delta \frac{D\vec{u}}{Dt} = -\vec{\nabla} p + \delta \vec{g}; \quad \vec{\nabla} \cdot \vec{u} = 0 \tag{7.18}$$

This is the so-called *Euler equation* for the motion of a *non-viscous*, incompressible fluid. It turns out, however, that viscous forces in a colloidal suspension are important, if not dominating, so Eq. (7.18) must be extended with the viscous stress on the blob. This stress is related to viscous transport of momentum as can be explained with reference to the sliding fluid layers in Fig. 7.1.

Shear forces. For the case of the plane-parallel flow in Fig. 7.1 the flow field is described by:

$$\vec{u} = [u(y), 0, 0]; \quad \vec{\nabla} \cdot \vec{u} = \frac{\partial u(y)}{\partial x} = 0 \tag{7.19}$$

Recall from Chap. 5 that $\vec{\nabla} \cdot \vec{u} = 0$ is the condition for steady, time-independent flow of an incompressible fluid, implying here that the velocity component in the x-direction depends only on y. The momentum $\vec{P}$ carried by this flow field is

$$\vec{P} = [P_x, P_y, P_z] = [m \, u(y), 0, 0] \tag{7.20}$$

Here m is the mass of a quantity of fluid moving at speed $u(y)$. For the change in momentum we can write, in general, see Eq. (7.5):

$$d\vec{P} = \frac{\partial \vec{P}}{\partial t} dt + \frac{\partial \vec{P}}{\partial x} dx + \frac{\partial \vec{P}}{\partial y} dy + \frac{\partial \vec{P}}{\partial z} dz \tag{7.21}$$

In the present case all partial derivatives are zero, except the one with respect to y, and (7.21) simplifies to:

$$\frac{dP_x}{dt} = \frac{\partial P_x}{\partial y}\frac{dy}{dt} = \frac{\partial mu(y)}{\partial y}\frac{dy}{dt} \tag{7.22}$$

This equation describes a shear force in the x-direction acting on a fluid area at position y. In other words: x-directed momentum is transported in the y-direction: a slowly moving fluid layer at y receives momentum from a faster layer at $y + dy$. The *shear stress* σ is defined as the shear force divided by the area A on which it is working:

$$\sigma = \frac{dy}{A\,dt}\frac{\partial mu(y)}{\partial y} = L\frac{dy}{dt}\frac{\partial \delta u(y)}{\partial y} \tag{7.23}$$

Here δ is the mass density of the fluid, and L a distance over which x-momentum diffuses in the y-direction. The *kinematic* viscosity v is defined as the ratio of the stress σ and the gradient in the moment density:

$$\sigma = v\frac{\partial \delta u(y)}{\partial y}; \quad v = \frac{\eta}{\delta}, \tag{7.24}$$

where η is the *shear* viscosity of fluid. Before proceeding to Newton's law in (7.28) we look at the analogy between (7.24) and Fick's law for the diffusion flux of particles.

Momentum diffusion. The flux of momentum is proportional to the gradient in momentum concentration, just as the flux of particles is proportional to the gradient in particle concentration. The coefficient v in (7.24) is indeed a diffusion coefficient with dimension m^2/s—consistent also with (7.23). Furthermore, when the momentum diffuses a distance L we find from (7.23) and (7.24)

$$L\int_0^L dy = v\int_0^t dt, \tag{7.25}$$

which leads to an instance of Einstein's law for quadratic displacement from Chap. 6:

$$L^2 \sim vt, \tag{7.26}$$

where v is a 'momentum diffusion coefficient'. Thus the time τ_H needed for momentum to propagate a distance of order sphere radius R is:

$$\tau_H \sim \frac{R^2}{v} = \frac{R^2\delta}{\eta} \tag{7.27}$$

This is the hydrodynamic time scale already anticipated in Chap. 4.

Fig. 7.2 A flow field exerts viscous forces on a volume element

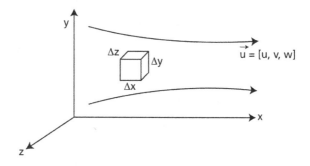

Newton's viscosity law. The analogy between diffusion of momentum and diffusion of particles is somewhat veiled when for a liquid of constant mass density δ, Eq. (7.24) is rewritten to its usual form:

$$\sigma_{xy} = \eta \frac{\partial u(y)}{\partial y} \tag{7.28}$$

which is known as *Newton's viscosity law*. Though we obtained this law from the simple flow of Eq. (7.19), it is valid for the general flow pattern described by Eq. (7.1). Note the convention of indices: σ_{xy} is the x-directed shear stress on a fluid layer at y. Alternatively one can say that σ_{xy} is the flux of x-momentum in the y-direction.

In rectangular coordinates nine stress components as in Eq. (7.28) may be written down: σ_{yx}, σ_{yy}, σ_{yz}, etc. Consider a volume element in a flow field (Fig. 7.2). The stress component σ_{yx} works on the surface elements $\Delta x \Delta y$ so the corresponding viscous force component on the volume element is:

$$\Delta x \Delta z \left[\sigma_{xy}(y + \Delta y) - \sigma_{xy}(y)\right] \tag{7.29}$$

Per unit volume this force equals

$$\frac{\partial \sigma_{xy}}{\partial y} = \eta \frac{\partial^2 u}{\partial y^2} \tag{7.30}$$

Considering all nine stress components the total viscous force per volume is

$$\eta \left(\frac{\partial^2}{\partial x^2} + \frac{\partial^2}{\partial y^2} + \frac{\partial^2}{\partial z^2}\right)[u, v, w] = \eta \nabla^2 \vec{u} \tag{7.31}$$

The Laplacian ∇^2 is here expressed in rectangular co-ordinates x, y, z. The more general form of the Laplacian in the Appendix B also applies to curvi-linear co-ordinates. Adding Eq. (7.31) to the Euler equation (7.18) we find:

$$\delta \frac{D\vec{u}}{Dt} = -\vec{\nabla} p + \delta \vec{g} + \eta \nabla^2 \vec{u}; \quad \nabla \cdot \vec{u} = 0 \tag{7.32}$$

Fig. 7.3 Left: example of turbulent water flow into a glass. Right: viscous flow of honey. Note the well-defined honey flow field, in contrast to the more chaotic pattern in the water. To undergo what bacteria experience in water, one should take a low-Reynolds number bath in a large tank of honey

This is the *Navier-Stokes* equation for an incompressible fluid with constant viscosity η and constant mass density δ. No single general solution of (7.32) exists; such a single solution would describe all possible flow patterns $\vec{u}$ and would generate, for example, both flow patterns in Fig. 7.3. A serious difficulty arises in particular at high velocities when chaotic turbulent flow occurs, in which the velocity and pressure are no longer unique functions of space and time coordinates. Turbulent flow can be observed everywhere (Fig. 7.3) and common as it may be, its theoretical description is extremely complicated.

A sufficiently slow, steady flow is stable against the occurrence of turbulence. For a steady flow $\partial u / \partial t = 0$ in the material derivative D/Dt of Eq. (7.10), so the equation of motion becomes:

$$\delta(\vec{u} \cdot \vec{\nabla})\vec{u} = -\vec{\nabla} p + \delta\vec{g} + \eta\nabla^2\vec{u} \qquad (7.33)$$

A further simplification involves the omission of the inertial terms $\delta(\vec{u} \cdot \vec{\nabla})\vec{u}$ resulting in the so-called *Stokes* equation. This simplification is fortunately justified for the small-scale flow patterns involving colloidal particles for reasons explained in the next section.

7.3   Stokes Flow

The example of flow in a curved tube in Fig. 7.4 illustrates the physical meaning of the various terms in Eq. (7.33). The hydrostatic pressure $\delta\vec{g}z$ which has a gradient in the vertical z-direction, induces liquid flow $\vec{u}$ and the concomitant pressure distribution p.

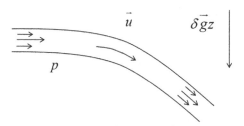

Fig. 7.4 Flow through a curved tube of a liquid with mass density δ, described by Eq. (7.33). Changes in flow velocity are resisted by the liquid's inertia according to the LHS of (7.33). Velocity gradients lead to energy dissipation quantified by the viscous term at the RHS of (7.33)

If the flow is steady nothing changes in time in any fixed volume element. Following a streamline we notice that the flow velocity changes direction which is 'resisted' by inertia: the $\delta(\vec{u} \cdot \vec{\nabla})\vec{u}$ term in Eq. (7.33). In a cross section of the tube a velocity gradient is present leading to the viscous term $\eta\nabla^2\vec{u}$.

The Reynolds number. The relative contributions from inertia and viscosity are estimated from a dimensionless parameter known as the *Reynolds number Re*:

$$\mathrm{Re} = \delta\frac{UL}{\eta}, \tag{7.34}$$

in which U is a typical flow speed and L a characteristic length scale of the flow. In Fig. 7.4, for example, U is the average flow velocity and L the diameter or length of the tube. The viscous term will dominate inertia if $Re \ll 1$.

To understand the origin of the inequality $Re \ll 1$ we note that derivatives of velocity components such as $\partial u/\partial x$ (describing inertia effects) are of order U/L, whereas second derivatives (designating viscous effects) are of order U/L^2. This gives the order of magnitude estimates:

$$\left|(\vec{u} \cdot \vec{\nabla})\vec{u}\right| \sim U^2/L; \quad \left|\nabla^2\vec{u}\right| \sim U/L^2 \tag{7.35}$$

The ratio of the two terms in Eq. (7.33) is therefore

$$\frac{\text{inertial term}}{\text{viscous term}} = \frac{\delta(\vec{u} \cdot \vec{\nabla})\vec{u}}{\eta\nabla^2\vec{u}} \sim \frac{\delta U^2/L}{\eta U/L^2} = \delta\frac{UL}{\eta}, \tag{7.36}$$

which equals the Reynolds number in Eq. (7.34). For colloidal particles typical values of UL are small enough to ensure that $Re \ll 1$ (Exercise 1). Then we may neglect the inertia term in Eq. (7.33) to obtain (for $\vec{g} = 0$):

$$0 = -\vec{\nabla}p + \eta\nabla^2\vec{u}; \quad \nabla \cdot \vec{u} = 0 \tag{7.37}$$

This is the *creeping flow* equation for purely viscous flow of an incompressible, Newtonian fluid, also known as *Stokes flow*.

7.4   On Magnitude

The term 'creeping flow' does not necessarily imply that flow is sluggish; it denotes a flow rate U which is small enough such that $Re \ll 1$. On the colloidal length scale (L in the submicron range) this rate may be actually quite high: micron colloids may settle under gravity or in a centrifuge at a rate of several times their diameter per second.

Small enough creatures such as bacteria live at low Reynolds numbers and, hence, experience water as a viscous fluid, in which they are completely unaware of their mass. We are large enough to suffer from inertia in a swimming pool, and will only have the Stokes flow experience when immersed in a bath of a very viscous fluid such as syrup or the honey from Fig. 7.3.

Reversible flow. One surprising feature of creeping flow is its reversibility which can be demonstrated (in a famous experiment by G. I. Taylor) as follows. Fill the gap of a *Couette* geometry (two concentric cylinders) with viscous oil and insert a dyed blob of oil with a syringe. The blob is sheared by slowly rotating one cylinder a few revolutions. However, if the cylinder is rotated back to its original position, the blob will return almost exactly to its original shape. The reversibility of purely viscous flow has an interesting biological consequence: if an animal is small enough to live at low Reynolds numbers, is has to battle this reversibility in order to move. A microbe, for example, trying to swim by flapping its tail to and fro makes no progress, because the effect of one flap is undone by the opposite flap. We are all living evidence of the fact that spermatozoa use their tail in a more efficient manner to swim in viscous bio fluids.

Not our scale. So what sort of hydrodynamics and external forces dominate the live of a creature, is a matter of scale. We end here with a memorable quote from the classical scholar and biologist D'Arcy Wentworth Thompson (1860–1948) who finalizes his book chapter[2] *On Magnitude* as follows:

> Life has a range of magnitude [....] wide enough to include three such discrepant conditions as those in which a man, an insect and a bacillus have their being and play their several roles. Man is ruled by gravitation, and rests on mother earth. A water-beetle finds the surface of a pool a matter of life and death, a perilous entanglement or an indispensable support. In a third world, where the bacillus lives, the resistance defined by Stokes's law, the molecular shocks of the Brownian motion, doubtless also the electric charges of the ionized medium, make up the physical environment and have their potent and immediate influence on the organism. The predominant factors are no longer those of our scale; we have come to the edge of a world of which we have no experience, and where all our preconceptions must be recast.

---

[2]D'Arcy Wentworth Thompson, *On Growth and Form*, Dover, 1992—the unabridged republication of the work published by Cambridge University Press, 1942.

Exercises

7.1 Verify that the Reynolds number Re is dimensionless.

7.2 Estimate for the Particle Quartet from Table 1.1 the Reynolds number for sedimentation due to gravity in water.

7.3 (a) Suppose you swim at a speed of $U = 2$ m/s; assuming a frontal area of 0.1 m^2, how large is Re? (b) To maintain a constant swim speed you have to invest energy because of (1) viscous friction on your body and (2) displacement of water. Argue which factor is the most important.

7.4 A sphere with diameter d is pulled out (at constant volume) to a thin rod with aspect ratio $L/D = 30$. Argue by which factor Re goes up or down.

7.5 (a) A force F moves a very large flat plate with constant speed $u(D)$, at a distance $y = D$ from a parallel wall in water. (a) Derive the velocity profile $u(y)$ from the Stokes equations (7.37), and give an expression for the average flow velocity $<u>$.

(b) Show that $u(y)$ is a stream function, and that it satisfies the continuity equation.

(c) Suppose $u(D) = 1$ mm s^{-1}; $D = 1$ mm and $\eta = 10^{-3}$ Pa s.
How large is F (per unit area)?

References

For more extensive treatments of the Navier-Stokes equation and Stokes flow see R. Bird, W. Stewart and E. Lightfoot, *Transport Phenomena* (New York, Wiley, 2002), and D.J. Acheson, *Elementary Fluid Dynamics* (Oxford, Clarendon Press, 1992).

More discussion on the viscosity dominated world of micro-organisms and colloids is given in: E.M. Purcell, *Life at low Reynolds number*, Am. J. Physics 45 (1977) 3–11.

Chapter 8
Flow Past Spheres and Simple Geometries

Flow problems in colloidal systems either concern flow in channels or flow around submerged particles. Channel flow, also known as *Poiseuille* flow, occurs for example for colloidal dispersions in a capillary for electrophoresis or electro-osmosis and for fluids in the tube or double-cylinder (Couette) geometry of a viscosity meter. Flow around particles arises for colloids undergoing sedimentation or Brownian motion. Though our primary goal is the hydrodynamic friction factor for flow past a sphere, we will first, as a warming-up, solve the Stokes equations for viscous flow in channels with a simple geometry.

8.1 Slits and Tubes—and Darcy's Law

Flow between flat plates. We start with a slit in the form of two flat parallel plates as in Fig. 8.1 at a distance d. There is only fluid motion in the x-direction so the flow velocity field in this geometry has the form

$$\vec{u} = [u(y), 0, 0] \tag{8.1}$$

This *plane parallel* flow satisfies $\vec{\nabla} \cdot \vec{u} = 0$, because the velocity component $u(y)$ is independent of x. For this flow pattern the Stokes equation is:

$$\frac{\partial p}{\partial x} = \eta \frac{\partial^2 u}{\partial y^2}, \quad \frac{\partial p}{\partial y} = \frac{\partial p}{\partial z} = 0 \tag{8.2}$$

Since $\partial p / \partial x$ is constant in the y-direction, $\partial^2 u / \partial y^2 = $ constant so u must be a quadratic function of y. Integrating (8.2) twice indeed yields the parabola:

$$u = \frac{1}{2\eta} \frac{dp}{dx} y(y - d), \tag{8.3}$$

© Springer Nature Switzerland AG 2018
A. P. Philipse, *Brownian Motion*, Undergraduate Lecture Notes in Physics,
https://doi.org/10.1007/978-3-319-98053-9_8

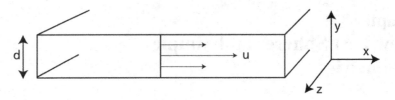

Fig. 8.1 Viscous flow between two parallel plates, with the parabolic liquid flow velocity profile given by Eq. (8.3)

a solution which satisfies the *stick boundary* or no-slip condition

$$u(y = 0) = u(y = d) = 0, \tag{8.4}$$

stating that at the surface of the plates the fluid is in rest relative to the plates; stick and slip boundaries are further addressed in Sect. 8.4. The flow velocity averaged over the volume in the gap between the two plates is:

$$< u > = \frac{\int_0^d \int_0^L u \, dx \, dy}{\int_0^d \int_0^L dx \, dy} = \frac{d^2}{12\eta} \frac{\Delta P}{L} \tag{8.5}$$

Here ΔP is the total pressure drop going from $x = 0$ to $x = L$. Apart from a numerical constant, Eq. (8.5) also arises from a dimensionless form of the Stokes equation as follows. Suppose $<u>$ is the velocity averaged in the y-direction over a length d, and ΔP is the pressure drop over length L. Introducing dimensionless parameters $\tilde{p}$, $\tilde{x}$, $\tilde{u}$ and $\tilde{y}$ defined as:

$$p = \tilde{p}\Delta P, \quad x = \tilde{x}L, \quad u = \tilde{u} < u >, \quad y = \tilde{y}d, \tag{8.6}$$

the Stokes equation (8.2) becomes:

$$\frac{\Delta P}{L} \frac{\partial \tilde{p}}{\partial \tilde{x}} = \frac{\eta}{d^2} < u > \frac{\partial^2 \tilde{u}}{\partial \tilde{y}^2} \tag{8.7}$$

Since the two derivatives only contain dimension-less quantities we can write:

$$< u > = \frac{d^2}{\eta} \frac{\Delta P}{L} \times \text{numerical factor} \tag{8.8}$$

So we can expect that Stokes flow in another geometry will have an average fluid velocity with the same functional form as the result for flat-plates in Eq. (8.5). We will verify this expectation for flow in a capillary with a circular cross-section.

Flow in a cylinder. The flow velocity $\vec{u}$ for *axial* flow in a tube of radius R, parallel to the z-axis (see Fig. 8.2) has the components:

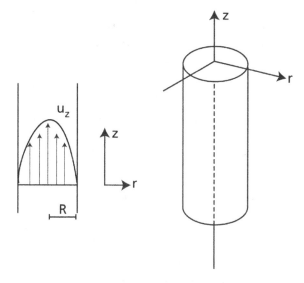

Fig. 8.2 Axial flow in a straight tube; the velocity profile is given by Eq. (8.15)

$$\vec{u} = \left[u_r, u_\theta, u_z\right] = \left[0, 0, u_z(r)\right] \tag{8.9}$$

Here we use cylindrical coordinates (r, θ, z); note that velocity components in the θ and r direction are zero. The Stokes equation for this type of flow is:

$$\frac{1}{\eta}\frac{\partial p}{\partial z} = \frac{1}{r}\frac{\partial}{\partial r}\left(r\frac{\partial}{\partial r}u_z\right) + \frac{\partial^2}{\partial z^2}u_z; \quad \vec{\nabla}\cdot\vec{u} = \frac{\partial u_z}{\partial z} = 0 \tag{8.10}$$

The zero-divergence of the flow velocity implies that u_z is constant in the z-direction, as expected for an incompressible liquid. Clearly then also $\partial^2 u_z/\partial z^2$ is zero:

$$\frac{1}{\eta}\frac{dp}{dz} = \frac{1}{r}\frac{d}{dr}\left(r\frac{d}{dr}u_z\right) \tag{8.11}$$

One integration yields the velocity gradient:

$$\frac{du_z}{dr} = \frac{1}{2\eta}\frac{dp}{dz}r + \frac{C}{r}; \quad C = 0 \tag{8.12}$$

The constant C must be zero, because otherwise this gradient is infinite at $r = 0$, which would imply an infinite stress. From this velocity gradient we obtain the viscous stress via Newton's viscosity law, derived in Chap. 7:

$$\sigma_{rz} = -\eta\frac{du_z}{dr} \tag{8.13}$$

This viscous stress can be employed to calculate the total viscous force on the inner wall of the tube (Exercise 8.2). The second boundary condition, in addition to the absence of an infinite stress at the center axis of the cylinder at $r = 0$, is the no-slip boundary at the wall of the tube:

$$\sigma_{rz}(r = 0) = 0; \quad u_z(r = R) = 0 \tag{8.14}$$

The solution for Eq. (8.12) which also satisfies this second condition is:

$$u_z = \frac{1}{4\eta}\frac{dp}{dz}(r^2 - R^2), \tag{8.15}$$

The average velocity in the tube is:

$$<u> = \frac{\int_0^L \int_0^R u_z r \, dr \, dz}{\int_0^L \int_0^R r \, dr \, dz} = \frac{R^2}{8\eta}\frac{\Delta P}{L} \tag{8.16}$$

The similarity to the flat plate result in Eq. (8.5) is clear; the different geometry only changes the numerical factor in Eq. (8.8). Note that the volume rate of flow Q

$$Q = <u> \pi R^2 = \frac{\pi R^4}{8\eta}\frac{\Delta P}{L} \tag{8.17}$$

strongly depends on the tube radius R. This result, called the *Hagen-Poiseuille law*, is the basis of viscosity measurements on (Newtonian) dispersions from flow rates in a tube. The R^4 scaling in the Hagen-Poiseuille equation also explains why only minor clogging of arteries may significantly raise the blood pressure (Exercise 8.7).

Tube friction factor. For the circular tube we can define a friction factor f_C as the ratio between the net force applied on the liquid in the tube, and the average liquid speed in (8.16). The result is:

$$f_C = \frac{\Delta P \pi R^2}{<u>} = 8\pi\eta L \tag{8.18}$$

The tube's friction factor differs by only a factor 4/3 from the friction factor $6\pi\eta L$ of a sphere with radius L.

Darcy's law. We note here in passing Darcy's law for viscous flow in a porous medium which states that the average flow velocity $<u>$ is proportional to the average pressure gradient that drives the flow:

$$<u> = -\frac{k}{\eta}<\vec{\nabla}p> \tag{8.19}$$

Here k is the so-called *liquid permeability* of the porous medium. For one-dimensional flow in a medium of length L Darcy's law becomes:

$$< u > = -\frac{k}{\eta}\frac{\Delta P}{L},\tag{8.20}$$

where ΔP is the total pressure drop over the length L. The permeability depends on the geometry of the medium and can only be calculated for simple cases. We have, in fact, already made this calculation for flow in tubes and between parallel plates. So Darcy's law in its integrated form (8.20) is just an instance of Eq. (8.8), with all geometrical details 'hidden' in the numerical factor. The liquid permeability k is proportional to the square of a typical 'pore diameter' d, a proportionality which also holds for more complicated pore geometries.[1]

8.2 Friction Factor of a Rotating Sphere

Viscous flow past a sphere. In the previous examples of flow through a channel the total viscous force on the inner wall of the channel equals the external force (i.e. the pressure drop Δp) which drives the flow (Exercise 2). The analogous force balance for viscous flow past a colloidal particle defines the Stokes friction factor f.

$$K = fu\tag{8.21}$$

Here u is the constant liquid velocity relative to the particle and fu is the total viscous force which balances the external force on the colloid. For a rotating particle we have to read K as a torque and u as an angular velocity, see Eq. (8.22). We will first calculate the rotational friction factor which involves a simpler flow field than for translational friction.

Rotating Stokes flow. Consider a solid sphere of radius R that slowly rotates at a constant angular velocity Ω around the z-axis in a large volume of quiescent fluid, see Fig. 8.3. We ask for the torque T_z required to maintain the sphere rotation, which defines the rotational friction factor f_r via:

$$T_z = f_r\Omega\tag{8.22}$$

In terms of spherical coordinates (r, θ, ϕ) the flow field near the sphere will be of the form:

$$\vec{u} = \left[u_r, u_\theta, u_\phi\right] = \left[0, 0, u_\phi(r, \theta)\right],\tag{8.23}$$

---

[1]Examples are the tortuous pore spaces in particle packings, see: D. M. E. Thies-Weesie and A. Philipse, J. Colloid and Interface Science **162**, 470–480 (1994).

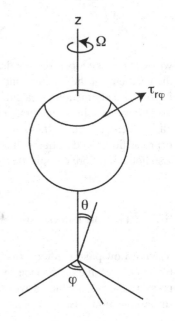

Fig. 8.3 A solid sphere rotates at constant angular velocity. The sphere exerts a stress $\tau_{r\phi}$, given by Eq. (8.31), in the ϕ—direction on the fluid

This field is symmetric about the z-axis of rotation so there is no dependence on the angle ϕ. The pressure p will be of the form $p = p(r, \theta)$ again without any ϕ—dependence. For the sphere the Stokes equation therefore adopts the form:

$$\eta\left[\nabla^2 \vec{u}\right]_\phi = \left[\vec{\nabla} p\right]_\phi = 0 \tag{8.24}$$

From the ϕ-component of $\nabla^2\vec{u}$, the Laplacian for spherical coordinates (see Appendix B), we only need the derivatives that depend on θ and r. Thus the Stokes equation becomes:

$$0 = \frac{1}{r^2}\frac{\partial}{\partial r}\left(r^2\frac{\partial u_\phi}{\partial r}\right) + \frac{1}{r^2}\frac{\partial}{\partial \theta}\left(\frac{1}{\sin\theta}\frac{\partial}{\partial \theta}\left(u_\phi \sin\theta\right)\right) \tag{8.25}$$

Flow field. Since there is no distinction between 'up' and 'down' in the flow field, the substitution $\theta \to \pi - \theta$ should not change the flow velocity, which suggests that u_ϕ is proportional to $\sin\theta = \sin(\pi -\theta)$. Since u_ϕ only depends on θ and r we choose as a trial solution for the flow field:

$$u_\phi(r, \theta) = f(r)\sin\theta \tag{8.26}$$

Insertion of this trial solution in Eq. (8.25) leads to the following differential equation for $f(r)$:

$$\frac{d}{dr}\left(r^2\frac{df}{dr}\right) - 2f = 0 \tag{8.27}$$

Here, the trial solution is $f = r^n$ which on substitution in Eq. (8.27) gives $n = 1$ and $n = -2$. Thus the flow field in (8.26) becomes

$$u_\phi(r, \theta) = \left(C_1 r + \frac{C_2}{r^2}\right)\sin\theta \tag{8.28}$$

To determine the constants C_1 and C_2 we note that at infinity all velocity components are zero, and that on the sphere's surface the liquid rotates with the same velocity as the sphere (stick boundary):

$$\begin{aligned} u_\phi &\to 0 &&\text{as } r \to \infty \\ u_\phi &\to R\Omega \sin\theta \text{ as } r = R \end{aligned} \tag{8.29}$$

Here $R\sin\theta$ is the shortest distance of a point on the surface of the sphere to the rotation axis z. The point traverses a circle with circumference $2\pi R\sin\theta$ with velocity $\Omega R\sin\theta$. Application of these boundary conditions to (8.28) shows that $C_1 = 0$ and $C_2 = \Omega R^3$. Therefore the final expression for the flow field induced by the rotating sphere is:

$$u_\phi = \Omega R \sin\theta\left(\frac{R}{r}\right)^2 = u_\phi(r = R)\left(\frac{R}{r}\right)^2 \tag{8.30}$$

Torque on the sphere. The relevant component of the stress is (see Appendix B):

$$\tau_{r\phi} = -\eta r\frac{\partial}{\partial r}\left(\frac{u_\phi}{r}\right) \tag{8.31}$$

To find the total torque we need to integrate the tangential force $\tau_{r\phi}(r = R)dS$ exerted on the fluid by a solid surface element dS, multiplying each element by its lever arm $R\sin\theta$ with respect to the rotation axis:

$$\begin{aligned} T_z &= \int \tau_{r\phi}(r = R)R\sin\theta dS \\ &= \int_0^{2\pi}\int_0^\pi (3\eta\Omega\sin\theta)(R\sin\theta)R^2\sin\theta d\theta d\phi = 6\pi\eta R^3\Omega\int_0^\pi \sin^3\theta d\theta \\ &= 8\pi\eta R^3\Omega \end{aligned} \tag{8.32}$$

By comparison with Eq. (8.22) we conclude that the rotational friction factor for a sphere in a pure viscous fluid is given by:

$$f_r = 8\pi \eta R^3, \tag{8.33}$$

so the rotational diffusion coefficient of the sphere equals:

$$D_r = \frac{kT}{8\pi \eta R^3} \tag{8.34}$$

Perhaps we expected the rotational friction factor to be proportional to R^2 namely the surface area between the rotating sphere and the surrounding fluid. Indeed, the total viscous force scales with R^2 but in (8.32) we need a viscous *torque* which entails an additional R-term in the lever arm leading to the scaling $f_r \sim R^3$.

Sphere in a cavity. The rotational friction factor in (8.33) has been derived for a sphere in an unbounded fluid, far away from a confining wall or other spheres. For one particular confinement the friction factor can be easily corrected. Suppose the sphere is rotating in a spherical cavity with radius $(1+\delta)R$, with $\delta \geq 0$. If the cavity represents a stick boundary which is at rest with respect to the rotation axis, we have instead of (8.29) the boundary conditions:

$$\begin{aligned} u_\phi &\to 0 && \text{as } r \to (1+\delta)R \\ u_\phi &\to R\Omega \sin\theta && \text{as } r = R \end{aligned} \tag{8.35}$$

Evaluating the constants C_1 and C_2 in Eq. (8.28), we find the flow field:

$$\frac{u_\phi}{\Omega R \sin\theta} = \frac{(1+\delta)^3 (R/r)^2 - (r/R)}{(1+\delta)^3 - 1} \tag{8.36}$$

Substitution of this flow field in the stress component $\tau_{r\phi}$ from Eq. (8.31) we again obtain the torque T_z, to find eventually for the rotational friction factor:

$$f_r = 8\pi \eta R^3 \frac{(1+\delta)^3}{(1+\delta)^3 - 1} \tag{8.37}$$

Note that this result reduces, as it should, to $8\pi \eta R^3$ in the limit $\delta \to \infty$ of an unbounded fluid. The rotational diffusion coefficient of the sphere in the cavity is accordingly:

$$D_r = \frac{kT}{8\pi \eta R^3} \left[1 - (1+\delta)^{-3} \right] \tag{8.38}$$

This simple extension of the rotational diffusion coefficient is also relevant for a sphere rotating in a complex fluid (such as a polymer solution) instead of a continuous solvent.[2]

---

[2] See for example: G. H. Koenderink et al., Rotational and translational diffusion of fluorocarbon tracer spheres in semi-dilute xanthan solutions, Phys. Rev. E (2004) **69**, 021804-1–12.

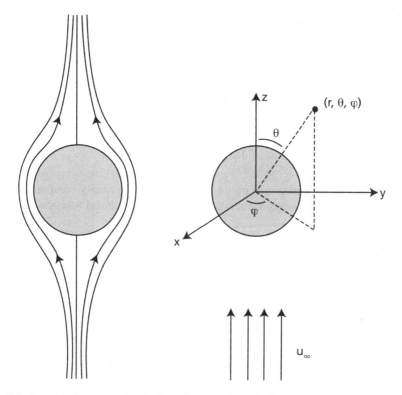

Fig. 8.4 Creeping flow past a fixed sphere. Far away from the fluid has a uniform speed u_∞; in the vicinity of the sphere, the fluid velocity profile is given by Eq. (8.60): A point in the fluid is specified by distance r from the origin, polar angle θ and a zimuthal angle φ

8.3   The Translational Friction Factor

We now determine the solution of the Stokes equations for creeping flow past a translating sphere. The non-rotating sphere in Fig. 8.4 is fixed in a fluid which has a uniform speed u_∞ far away from the sphere. Using spherical coordinates (r, θ, ϕ), the flow field near the sphere is of the form:

$$\vec{u} = [u_r(r, \theta), u_\theta(r, \theta), 0] \qquad (8.39)$$

This is a two-dimensional 'axisymmetric' flow: the fluid approaches from the z-direction so if we observe the flow in a plane perpendicular to the z-axis at a fixed distance r the pattern is the same for every angle ϕ; the velocity component u_ϕ in Eq. (8.39) is zero. In contrast to Poiseuille flow in a straight tube streamlines are now curved, which makes the creeping flow equation more difficult to solve directly for the components of $\vec{u}$. One option is to simplify the Stokes equation by rewriting it in terms of a stream function ψ instead of $\vec{u}$.

Stokes stream function. A stream function has a constant value along a stream-line. According to Eq. (7.11) ψ is a stream function if

$$(\vec{u} \cdot \vec{\nabla})\psi = 0 \tag{8.40}$$

In the case of plane parallel flow in Eq. (8.1), the velocity component $u(y)$ is itself a stream function:

$$(\vec{u} \cdot \vec{\nabla})u(y) = u\frac{\partial u(y)}{\partial x} = 0 \tag{8.41}$$

Along the curved streamlines past a sphere, however, velocity components are not constant. The components are actually derivatives of a stream function:

$$u_r = \frac{1}{r^2 \sin\theta} \frac{\partial \psi}{\partial \theta}, \quad u_\theta = -\frac{1}{r \sin\theta} \frac{\partial \psi}{\partial r} \tag{8.42}$$

These equations define the Stokes stream function ψ, which is indeed a stream function because

$$(\vec{u} \cdot \vec{\nabla})\psi = u_r \frac{\partial \psi}{\partial r} + \frac{u_\theta}{r} \frac{\partial \psi}{\partial \theta} = 0 \tag{8.43}$$

For the velocity components in Eq. (8.42) it is also the case that

$$\vec{\nabla}.\vec{u} = 0 \tag{8.44}$$

This is the trick of the Stokes stream function: if we find ψ, the velocity $\vec{u}$ in Eq. (8.39) immediately follows while we automatically satisfy $\vec{\nabla}.\vec{u} = 0$. So from the creeping flow Eq. (7.37) we only need:

$$\vec{\nabla} p = \eta \nabla^2 \vec{u} = -\eta \vec{\nabla} \times \left(\vec{\nabla} \times \vec{u}\right), \tag{8.45}$$

where we have substituted the Laplacian from equation B12 in Appendix B. The curl of fluid velocity field in terms of the stream function in Eq. (8.42) is the vector:

$$\vec{\nabla} \times \vec{u} = \left[0, 0, \frac{-1}{r \sin\theta} E^2 \psi\right], \tag{8.46}$$

where E^2 is the differential operator

$$E^2 = \frac{\partial^2}{\partial r^2} + \frac{\sin\theta}{r^2} \frac{\partial}{\partial \theta}\left(\frac{1}{\sin\theta} \frac{\partial}{\partial \theta}\right) \tag{8.47}$$

Substitution of (8.46) in Eq. (8.45) yields:

$$\frac{\partial p}{\partial r} = \frac{\eta}{r^2 \sin\theta} \frac{\partial}{\partial \theta} E^2 \psi; \quad \frac{1}{r} \frac{\partial p}{\partial \theta} = \frac{-\eta}{r \sin\theta} \frac{\partial}{\partial r} E^2 \psi \tag{8.48}$$

Next we note that the pressure $p = p(r, \theta)$ is a state function (dp is an exact differential). Then by definition the order of differentiation may be reversed:

$$\frac{\partial}{\partial \theta} \frac{\partial p}{\partial r} = \frac{\partial}{\partial r} \frac{\partial p}{\partial \theta} \tag{8.49}$$

The pressure can now be eliminated by combining (8.48) and (8.49) to obtain

$$E^2 (E^2 \psi) = 0, \tag{8.50}$$

which on substitution of (8.47) can be rewritten to:

$$\left[\frac{\partial^2}{\partial r^2} + \frac{\sin\theta}{r^2} \frac{\partial}{\partial \theta} \left(\frac{1}{\sin\theta} \frac{\partial}{\partial \theta} \right) \right]^2 \psi = 0 \tag{8.51}$$

This is the simplified version of the differential equation (8.45) for $\vec{u}$.

Trial function. We now have to guess a form of ψ which satisfies (8.51). A suitable form suggests itself by the 'infinity condition': at $r \to \infty$ the flow becomes uniform with speed u_∞ (see Fig. 8.4) in the z-direction:

$$u_r \sim u_\infty \cos\theta \quad \text{and} \quad u_\theta \sim -u_\infty \sin\theta, \quad \text{as } r \to \infty \tag{8.52}$$

For the stream function in Eq. (8.42) this implies:

$$\psi \sim \frac{1}{2} u_\infty r^2 \sin^2\theta, \quad \text{as } r \to \infty, \tag{8.53}$$

which suggests a solution to Eq. (8.51) of the form

$$\psi = f(r) \sin^2\theta \tag{8.54}$$

Substitution of this trial solution in Eq. (8.51) shows that $f(r)$ follows from the differential equation:

$$\left(\frac{d^2}{dr^2} - \frac{2}{r^2} \right)^2 f(r) = 0 \tag{8.55}$$

We now try solutions of the form $f = r^\alpha$, which indeed satisfy Eq. (8.55) provided that:

$$[(\alpha - 2)(\alpha - 3) - 2][\alpha(\alpha - 1) - 2] = 0, \tag{8.56}$$

which is the case for $\alpha = -1, 1, 2, 4$. Therefore:

$$f(r) = \frac{A}{r} + Br + Cr^2 + Dr^4,$$ (8.57)

where A, B, C and D are constants that are determined as follows. The condition of uniform flow at infinity in Eq. (8.53) can only be fulfilled if $C = \frac{1}{2} u_\infty$ and $D = 0$. The stick-boundary condition that u_θ and u_r in Eq. (8.42) are both zero on the sphere surface implies

$$-\frac{\partial \psi}{\partial r} = \frac{1}{r}\frac{\partial \psi}{\partial \theta} = 0 \quad \text{on } r = R,$$ (8.58)

which reduces to $(1/R)f(R) = f'(R) = 0$. This determines the constants $A = u_\infty R^3/4$ and $B = -3u_\infty R/4$. The Stokes stream function finally turns out to be:

$$\psi = \frac{1}{4}u_\infty\left(2r^2 + \frac{R^3}{r} - 3Rr\right)\sin^2\theta$$ (8.59)

Streamlines as sketched in Fig. 8.4 correspond to certain values of ψ. For example $\psi = 0$ generates the streamline which satisfies either $r = R$ or $\theta = 0$. Note in Fig. 8.4 that the flow pattern has 'for-after' symmetry: the streamlines will remain the same if the flow u_∞ is reversed. This is another example of the reversibility of creeping flow referred to at the end of Chap. 7.

Velocity profile. By substituting ψ we can compute the velocity components in Eq. (8.42):

$$\frac{u_r}{u_\infty} = \left[1 - \frac{3}{2}\left(\frac{R}{r}\right) + \frac{1}{2}\left(\frac{R}{r}\right)^3\right]\cos\theta$$

$$\frac{u_\theta}{u_\infty} = -\left[1 - \frac{3}{4}\left(\frac{R}{r}\right) - \frac{1}{4}\left(\frac{R}{r}\right)^3\right]\sin\theta$$ (8.60)

One striking feature of this velocity profile of a translating sphere is its long range due to the R/r term-compare, for example, the flow field around a rotating sphere in (8.30). So a diffusing or sedimenting colloidal sphere causes a disturbance of a uniform flow which extends over many sphere diameters. Therefore these solutions to the Stokes equation for a single sphere are only valid if the sphere is far away from other spheres or a wall.

The radial pressure gradient in Eq. (8.48) turns out to be:

$$\frac{\partial p}{\partial r} = 3u_\infty\eta Rr^{-3}\cos\theta$$ (8.61)

At infinity, the pressure in the uniform flow is p_∞:

$$p = p_\infty - \frac{3}{2}u_\infty\eta\frac{R}{r^2}\cos\theta$$ (8.62)

As could be expected, the pressure exceeds the bulk pressure p_∞ at the sphere side which receives the flow. For a colloidal sphere settling or diffusing in a liquid we are interested in the net force which is exerted on the sphere in Fig. 8.4. The relevant viscous stress component is

$$\tau_{r\theta} = -\eta r \frac{\partial}{\partial r}\left(\frac{u_\theta}{r}\right) \tag{8.63}$$

namely the stress tangential to the sphere's surface due to the velocity gradient perpendicular to the surface. For the velocity component u_θ in Eq. (8.60) we obtain for the stress on the surface of the sphere:

$$\tau_{r\theta} = \eta \frac{3}{2}\frac{u_\infty}{R}\sin\theta, \quad \text{at } r = R \tag{8.64}$$

Further, the pressure on the surface is:

$$p = p_\infty - \eta\frac{3u_\infty}{2R}\cos\theta, \quad \text{at } r = R \tag{8.65}$$

By symmetry the net force on the sphere will be oriented in the z-direction, parallel to the uniform flow (Fig. 8.4). The relevant components of p and $\tau_{r\theta}$ on the sphere surface are

$$t = \tau_{r\theta}\sin\theta - p\cos\theta = \frac{3\eta u_\infty}{2R} - p_\infty\cos\theta \tag{8.66}$$

The total force F on the sphere is the integral of t over the whole sphere surface:

$$F = \int_0^{2\pi}\int_0^\pi t R^2\sin\theta \, d\theta d\varphi = 6\pi\eta u_\infty R \tag{8.67}$$

Note that the term $p_\infty\cos\theta$ in Eq. (8.66) does not contribute to this total force, because the isotropic bulk pressure p_∞ can have no net effect on the sphere. Thus the stress in Eq. (8.66) on the sphere surface is everywhere the same, a surprising result in view of the velocity profile in Fig. 8.4. The proportionality factor between the uniform flow velocity u_∞ far away from the sphere and the drag force is:

$$f = 6\pi\eta R, \tag{8.68}$$

This is the Stokes friction factor for the translational motion of a sphere in a viscous fluid, valid for small Reynolds numbers and under the condition of a no-slip boundary between sphere surface and fluid.

Sphere settling. One application of the Stokes friction factor in (8.68), concerns a sedimenting sphere which accelerates downwards in a fluid until a constant velocity

U_0 is achieved. Then the drag force balances the weight of the sphere, corrected for buoyancy:

$$6\pi \eta U_0 R = \frac{4}{3}\pi R^3 (\delta_{\text{sphere}} - \delta_{\text{fluid}})g, \tag{8.69}$$

where δ is a mass density. The stationary sedimentation velocity

$$U_0 = \frac{2R^2}{9\eta}(\delta_{\text{sphere}} - \delta_{\text{fluid}})g, \tag{8.70}$$

also called the Stokes velocity, of course only applies if all assumptions underlying the Stokes friction factor ($Re \ll 1$, no-slip boundary) are justified. The same applies of course for the Stokes-Einstein diffusion coefficient for the sphere:

$$D = \frac{kT}{f} = \frac{kT}{6\pi \eta R} \tag{8.71}$$

One assumption has not been addressed explicitly, namely that the fluid surrounding the sphere is a continuum. For a colloidal micron-sphere diffusing or settling in a low-molecular solvent this is certainly the case: on its diffusive time scale the sphere experiences continuum hydrodynamics, see the discussion on time scales in Chap. 4. For an ion, however, one would expect that the continuum hypothesis fails. Nevertheless, the Stokes friction factor Eq. (8.68) is widely used in the Stokes-Einstein diffusion coefficient of small solute molecules, and in many cases appears to work well; see also the remark on ion diffusion at the end of Sect. 6.1.

8.4   Stick, Slip and the Lotus Sphere

The Stokes-Einstein diffusion coefficients for sphere translation and rotation are often applied without attention being paid to the underlying assumption of a stick boundary condition. The derivation of the Stokes friction factors relies on fluid mechanics of a sphere suspended in a continuum for which we can neglect all molecular details; the great benefit of the stick-boundary condition is that these details can also be disregarded for fluid that contacts the sphere. As a result the only fluid property that enters into the Stokes friction factors is the macroscopic viscosity η. Slip, however, leads in friction factors to an additional parameter β whose value is not a priori known, because when a solid surface and the adjacent fluid move at different speeds the resulting frictional force will differ from one solid-liquid combination to another.

The parameter β can be defined as the constant of proportionality between the tangential stress τ and the relative solid-liquid speed u, at the sphere surface:

$$\tau = \beta u, \quad \text{at } r = R \tag{8.72}$$

Note the analogy with the friction factor f in Eq. (8.21): f is a force per unit of speed and β is a *stress* per unit of speed. The dimension of β is accordingly a friction factor per unit area.

For the tangential velocity component u_ϕ on a rotating sphere boundary condition (8.72) adopts the form

$$\beta\left(R\Omega\sin\theta - u_\phi\right) = \tau_{r\phi} = -\eta r \frac{\partial}{\partial r}\left(\frac{u_\phi}{r}\right), \quad \text{at } r = R \tag{8.73}$$

bearing in mind that u_ϕ is the velocity relative to the z-axis of rotation, see also Fig. 8.3. The flow field is, compare for example Eq. (8.28):

$$u_\phi = \frac{C}{r^2}\sin\theta, \tag{8.74}$$

which on substitution in Eq. (8.73) determines the constant C with the result:

$$u_\phi = \Omega R\left(\frac{R}{r}\right)^2 \sin\theta\frac{1}{1+3S}; \quad S = \frac{\eta}{\beta R} \tag{8.75}$$

Here S is a dimensionless parameter, which measures the 'amount' of slip at the sphere surface. The term $1/1+3S$ multiplies the flow field from Eq. (8.30) and obviously also the total torque T_z on the sphere in Eq. (8.32), so the rotational friction factor and the rotational diffusion coefficient modify to:

$$f_r = \frac{8\pi\eta R^3}{1+3S}; \quad D_r = \frac{(1+3S)kT}{8\pi\eta R^3} \tag{8.76}$$

The stick boundary condition is the limit $S \to 0$, where we recover our earlier result $f = 8\pi\eta R^3$. For the translational friction factor one can show (Exercise 4) that the boundary condition (8.72) leads to:

$$f = 6\pi\eta R\left(\frac{1+2S}{1+3S}\right) \tag{8.77}$$

In the limit of a pure no-slip boundary we recover the familiar friction factor $f = 6\pi\eta R$.

The Lotus sphere. Imagine the fate of a super-hydrophobic[3] sphere in water. For a strongly water repelling sphere, the pure slip boundary condition $S \to \infty$ applies, *i.e.* the tangential stress on the sphere surface is zero. This condition—at first sight—reduces the Stokes factor in (8.76) to $f_r = 0$. However, Lotus sphere and its surrounding water very likely will be separated by a thin layer composed of water vapor and dissolved gas. In that case the sphere performs thermal rotations in a water

---

[3] Super-hydrophobicity is also known as the Lotus effect; the leaves of the Lotus plant are not wetted by water due to a porous surface structure that entraps air.

cavity lubricated by a gas, with a rotational diffusion coefficient that is significantly larger than for a hydrophilic sphere to which water sticks.

Sphere translation, in contrast to rotation, displaces liquid such that in absence of viscous stress still a significant friction factor remains; from (8.77) we find for a Lotus sphere:

$$f = 4\pi\eta R, \quad \text{for } S \to \infty \tag{8.78}$$

So we may conclude that for spheres in a continuous fluid the incorporation of slip effects in friction factors is possible on the basis of the plausible boundary condition of Eq. (8.72).

Exercises

8.1 Give the equation for the volume rate of flow Q for the flat plates in Fig. 6.1.

8.2 Sketch the profile of the viscous stress τ_{rz} and show that the total viscous force on the inner wall of the tube in Fig. 8.2 equals $\pi R^2 \Delta P$.

8.3 Derive the Stokes friction factor (per unit length) for the rotation of a very long cylinder with radius R around its long axis (cf. Fig. 8.3).

8.4 Equation (8.76) seems not to have been reported earlier. Formula (8.77), however, can be found in a somewhat different notation in J. Happel and H. Brenner, *Low Reynolds Number Hydrodynamics* (Englewood Cliffs, NJ: Prentice-Hall, 1965, pp 125–126). Verify that the formula is correct.

8.5 Re-examine the Poiseuille flow in simple geometries with a "pure-slip" boundary condition. Conclusion?

8.6 Calculate the gravitational settling rates of the Particle Quartet of Table 1.1.

8.7 A certain fatty deposit decreases the inner radius of a blood capillary from $R_1 = 5$ micron to $R_2 = 4$ micron. By which percentage should pressure p increase to keep transport of red blood cells at the same level as for a clean blood vessel?

References

The derivation of the translational Stokes friction factor is adapted from: D.J. Acheson, *Elementary Fluid Dynamics*, (Oxford, Clarendon Press, 1992).

The derivation by Stokes himself can be found in: G.C. Stokes, *Mathematical and Physical Papers*, Vol. III, (Cambridge, Cambridge University Press, 1901).

The treatment of the rotational friction factor is adapted from a classic textbook with many practical transport problems: R. Bird, W. Stewart and E. Lightfoot, *Transport Phenomena*, (New York, Wiley, 2002).

An entrance to literature on 'stick versus slip' is S. Granick, Y. Zhu and H. Lee, *Slippery questions about complex fluids flowing past solids*, Nature Mat. 2 (2003) 221.

This Chapter deals with friction factors in continuous fluids; for their applicability to Brownian motion in 'discrete' host solutions, see G.H. Koenderink *et al.*, *On the validity of Stokes-Einstein-Debye scaling in colloidal suspensions*, Faraday Discuss. 123 (2003) 129–136.

Chapter 9
Encounters of the Brownian Kind

9.1 Diffusion Versus Convection

In a quiescent solution, in absence of any particle convection due to an external field, Brownian motion is the only transport mechanism for (non-living) colloidal particles to encounter each other. To get an idea of the time scale involved we compute the time taken by a sphere of radius R to diffuse a mean-square-displacement equal to R^2—the configurational relaxation time τ_{CR} introduced in Chap. 4. For spheres in water at room temperature:

R (nm)	$\tau_{CR} \sim \eta R^3/kT$ (s)
10	2×10^{-7}
100	2×10^{-4}
1000	2×10^{-1}
10^4	2×10^2

Clearly, for small colloidal particles, Brownian motion on their own colloidal length scale is fairly rapid, whereas for radii much larger than a micron, diffusion is a hopelessly inefficient transport vehicle. Also for small, rapidly diffusing particles covering large distances is a slow process due to the square-root time dependence of diffusive displacements. For traversing large spaces convective transport must take over: we stir to homogenize coffee and milk rather than waiting for diffusion to do the mixing.

On the other hand, convection becomes an inadequate transport vehicle close to a surface, or in sufficiently narrow geometries, where the viscous drag is very large. So for small particles or molecules that have to react with a surface—or penetrate a biological cell—the profitable strategy is to cross large distances by convection, followed by Brownian motion for the final sub-micron steps. A typical biological

© Springer Nature Switzerland AG 2018
A. P. Philipse, *Brownian Motion*, Undergraduate Lecture Notes in Physics,
https://doi.org/10.1007/978-3-319-98053-9_9

Fig. 9.1 Poiseuille flow in a vertical capillary of length h and radius a, driven by pressure difference ΔP caused by the fluid weight. The question—answered in Sect. 9.1—is when the time needed for a diffusive displacement a of a colloid, equals the time needed for the fluid to propel the colloid the same distance

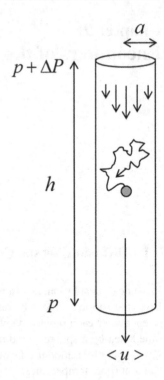

example is formed by viruses that are propelled by the flow of blood or air but that eventually have to arrive at suitable landing places on a new target cell via Brownian motion on a sub-micron scale.

Tubes and arteries. A simple case for examining the relative effects of diffusion and convection is provided by a colloid in the Poiseuille flow (Fig. 9.1) through a vertical capillary of radius a driven by a pressure difference

$$\Delta P = gh\delta \tag{9.1}$$

for a fluid with mass density δ in a capillary of length h; g is the gravitational acceleration. The time t_D needed for a colloid of radius R to diffuse a distance equal to the capillary radius a is of the order:

$$t_D \sim \frac{a^2}{2D} = \frac{3\pi \eta R a^2}{kT} \tag{9.2}$$

We should compare this to the time t_C taken by the colloid to travel the same distance by liquid convection. Ignoring the parabolic velocity profile (Fig. 9.1) we just employ the average flow velocity, derived earlier in Chap. 7:

$$<u> = \frac{a^2}{8\eta}\frac{\Delta P}{h} = \frac{a^2}{8\eta}g\delta \tag{9.3}$$

Hence the convection time is about:

$$t_C \sim \frac{8\eta}{ag\delta} \tag{9.4}$$

The ratio of diffusion time and convection time is therefore

$$\frac{t_D}{t_C} = a^3 R \frac{3\pi}{8} \frac{g\delta}{kT} \tag{9.5}$$

Note that this time ratio also follows from the Péclet number in the form

$$Pe = \frac{<u> a}{2D} \tag{9.6}$$

The capillary radius a_B for which the time ratio in (9.5) equals unity is about $a_B \approx 1.5$ and $a_B \approx 4$ μm for, respectively, the colloidal C-sphere ($R = 100$ nm) and the nano-sphere ($R = 5$ nm) from Table 1.1 (Exercise 9.6).

Blood vessels. These estimates for a_B show that for pore radii in the micron range there is a significant transport contribution from both Brownian motion and convection. This puts the radii of blood vessels into perspective. The narrow vessels referred to as 'blood capillaries' are located in body tissues and transport blood from arteries to the veins that bring blood back to the heart. Radii of these capillaries are in the range from about 2–5 μm, indicating that species of nano-meter size, let alone small molecules, have enough time to reach capillary walls during blood flow by diffusion.

9.2 Brownian Motion Towards a Spherical Absorber

Brownian collisions on a spherical target will be analyzed in more detail below because it captures the essential kinetics of many processes including coagulation of colloids, diffusional growth and diffusion-controlled chemical reactions. We consider a collection of Brownian spheres with radius R_j, diffusing in the vicinity of a target sphere with radius R_i centered at the origin (Fig. 9.2). The frequency of collisions of j-spheres on the target sphere can be found, following Smoluchowski, via a stationary diffusion model as follows.

Imagine that every j-sphere that hits the target sphere is somehow removed from the solution: the target act as an irreversible, spherical absorber such that the concentration of j-spheres near its surface is zero. Assuming that in a very large bulk far away from the central sphere the j-sphere density remains constant, a steady diffusion of j-spheres from the bulk to the target will be established. The continuity equation for the number concentration c_j of j-spheres is (see Chap. 5):

$$\frac{\partial c_j}{\partial t} = -\vec{\nabla} \cdot \vec{j}_d, \tag{9.7}$$

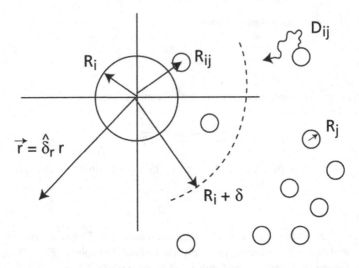

Fig. 9.2 Spheres j diffuse from a bulk (with number concentration $c_{j,\infty}$) at a distance $R_i + \delta$ towards a diffusing tracer sphere with radius R_i which acts as an infinite sink from which no j-sphere can escape. A practical instance is a porous sphere that irreversibly absorbs foul smelling j-molecules

which together with Fick's first law

$$\vec{j_d} = -D_{ij}\vec{\nabla}c_j \tag{9.8}$$

leads to the diffusion equation which we already encountered (in slightly different notation) in Chap. 5.

$$\frac{\partial c_j}{\partial t} = D_{ij}\nabla^2 c_j \tag{9.9}$$

Here D_{ij} is the diffusion coefficient of the j-spheres *relative* to the center of the target sphere which itself also exhibits Brownian motion.

Stationary diffusion. The concentration profile of j-spheres reaches a steady state when $\partial c_j/\partial t = 0$. Then (9.9) reduces to the Laplace equation:

$$\nabla^2 c_j = \frac{1}{r^2}\frac{\partial}{\partial r}\left(r^2\frac{\partial c_j}{\partial r}\right) = 0, \tag{9.10}$$

with the solution:

$$c_j(r) = \frac{A}{r} + B, \tag{9.11}$$

in which A and B are constants. The boundary conditions, see also Fig. 9.2, are a constant bulk concentration $c_{j,\infty}$ beyond some distance $R_i + \delta$ from the origin, and zero-concentration of j-spheres at the target surface:

$$c_j(r = R_i + \delta) = c_{j,\infty} \, ; \quad c_j(r = R_{ij}) = 0 \tag{9.12}$$

Note that the boundary condition of zero-concentration actually occurs at the collision distance R_{ij}, i.e. the center-to-center distance at which the i- and j-sphere touch. After evaluating A and B from the boundary conditions in (9.12), the steady-state profile turns out to be:

$$\frac{c_j(r)}{c_{j,\infty}} = \left(1 - \frac{R_{ij}}{r}\right)\left(\frac{R_i + \delta}{R_i + \delta - R_{ij}}\right) \tag{9.13}$$

The question now is which value we have to take for the 'diffusion-zone thickness' δ; in other words, where in Fig. 9.2 does the bulk begin? Fortunately, for a single target sphere in a sufficiently large container of j-particles, we do not have to specify δ any further than that it is much larger than R_{ij}. Thus for the steady-state profile we can take the limit

$$\lim_{\delta \to \infty} \frac{c_j(r)}{c_{j,\infty}} = 1 - \frac{R_{ij}}{r} \tag{9.14}$$

In what follows, we will only employ this concentration profile. We note here that the simple, asymptotic result in (9.14) is a fortunate consequence of Brownian motion in three-dimensional space: diffusion in a two-dimensional plane involves an undetermined δ (Exercise 1).

Stationary flux. The steady diffusion flux of j-spheres in radial direction (unit vector $\vec{\delta}_r$) towards the target follows from substitution of (9.14) in Fick's first law (9.8):

$$\vec{j}_d = -D_{ij}\vec{\delta}_r\frac{dc_j}{dr} = -D_{ij}\vec{\delta}_r\frac{d}{dr}c_{j,\infty}\left(1 - \frac{R_{ij}}{r}\right) = -\frac{\vec{\delta}_r}{r^2}D_{ij}R_{ij}c_{j,\infty} \tag{9.15}$$

This is a flux of particles (per unit area per second) which decreases with increasing r. However, because of mass conservation the *total* flux J through a spherical envelope of surface area $4\pi r^2$ must be independent of r. This independence also follows from the steady-state condition $\vec{\nabla} \cdot \vec{j}_d = 0$. Thus the collision-frequency of j-particles on the target sphere can be equated to this total flux, evaluated at $r = R_{ij}$:

$$J(j \to i) = 4\pi R_{ij}^2 \left|\vec{j}_d(r = R_{ij})\right| = 4\pi D_{ij}R_{ij}c_{j,\infty} \tag{9.16}$$

Next we note that for solid spheres the collision radius equals $R_{ij} = R_i + R_j$, and that for independently diffusing spheres their relative diffusion coefficient equals (Exercise 2):

$$D_{ij} = D_i + D_j = \frac{kT}{6\pi\eta}\left(\frac{1}{R_i} + \frac{1}{R_j}\right) \tag{9.17}$$

Consequently Eq. (9.16) becomes:

$$J(j \rightarrow i) = \frac{2kT}{3\eta}\left(2 + \frac{R_i}{R_j} + \frac{R_j}{R_i}\right)c_{j,\infty} \tag{9.18}$$

It is instructive to rewrite this collision frequency in terms of the volume fraction $\phi_j = (4/3)\pi R_j^3 c_{j,\infty}$ of j-spheres:

$$J(j \rightarrow i) = \left[2 + \frac{R_i}{R_j} + \frac{R_j}{R_i}\right]\frac{\phi_j}{2\pi\tau_{CR}}; \quad \tau_{CR} \sim \frac{\eta R_j^3}{kT} \tag{9.19}$$

This expression shows that, for given volume fraction, it is the configuration relaxation time τ_{CR} that determines the collision frequency of j-spheres on the target.

Effect of poly-dispersity. A remarkable feature of the diffusion flux $J(j \rightarrow i)$ is its minimum for spheres of identical size (Exercise 3):

$$J(j \rightarrow i) = \frac{2\phi_j}{\pi\tau_{CR}}; \quad R_i = R_j \tag{9.20}$$

In other words, for a given volume fraction, polydispersity always accelerates Brownian encounter frequencies in comparison to monodisperse spheres. The minimal value of (9.20) can be qualitatively understood by noting that if, in a monodisperse system, we shrink all spheres except the target sphere, the collision frequency increases due to the enhanced diffusion of the shrunk spheres. If, on the other hand, only the target sphere is reduced to a point-like particle, it will rattle around rapidly in a collection of static j-spheres which also increases the diffusion flux $J(j \rightarrow i)$.

To get an idea of the collision frequencies involved, imagine the fate of a target sphere with radius $R_i = 1$ μm immersed in an aqueous host dispersion with a particle volume fraction $\varphi_j = 0.01$. According to Eq. (9.19) the collision frequency on this target is in order of magnitude:

R_j (nm)	$J(j \rightarrow i)$ (s^{-1})
10	68×10^2
100	80×10^{-2}
1000	26×10^{-3}

Thus the micron-sized target is bombarded quite frantically by Brownian motion when the hosts are nano particles, whereas for an equal volume fraction of micron-sized hosts, the target has to wait for more than half a minute for the next Brownian encounter to take place.

9.3 Diffusional Sphere Growth

A Brownian particle may come into existence by clustering of solute molecules in a sufficiently concentrated solution—a process also referred to as 'precipitation'. In the initial stage of this precipitation process solute molecules form small, transient clusters by thermal fluctuations. Beyond a critical cluster size, particles irreversible grow by uptake of more solute molecules. The growth kinetics can—at least qualitatively—be explained in terms of the diffusion flux towards a target sphere discussed in the previous Sect. 9.2.

In the expression (9.16) for the collision frequency we substitute $D_j >> D_i$ and, consequently, also $R_{ij} \approx R_i$ so the collision frequency on a sphere is approximately that of small particles on a large, static target:

$$J(j \rightarrow i) \approx 4\pi D_j R_i c_{j,\infty} \qquad (9.21)$$

If each j-particle contributes a volume v_j to the volume V_i of the growing target we have for V_i the differential equation:

$$\frac{dV_i}{dt} = J(j \rightarrow i)v_j; \quad V_i = (4/3)\pi R_i^3 \qquad (9.22)$$

Substitution of (9.21) and integration yields for the radius R_i at time t:

$$R_i^2(t) - R_i^2(t_0) = 2D_j\phi_j(t - t_0); \quad \phi_j \approx c_{j,\infty}v_j \qquad (9.23)$$

The volume fraction φ_j is actually larger than the true volume fraction of j-particles because the volume contribution v_j to the growing sphere volume exceeds the j-sphere volume itself. Note in (9.23) the scaling $R \sim t^{1/2}$ which is characteristic for diffusion-controlled growth. The growth Eq. (9.23) is indeed an instance of Einstein's law for quadratic displacement, here in the form of a particle radius squared that grows linearly in time. The usual Stokes-Einstein single-particle diffusion coefficient D is replaced by an effective diffusion coefficient $D_j\varphi_j$.

Diffusion coefficients of small molecules or ions in water are typically on the order of $D_j \sim 10^{-5}$ cm^2 s^{-1} so for a volume fraction $\varphi_j = 0.01$, Eq. (9.23) predicts a sphere growth rate of about $dR^2/dt \approx 20$ μm^2/s. This is quite fast; growth of colloidal spheres by precipitation in a supersaturated solution is often much slower. Retarding factors include exhaustion of the bulk (decrease of φ_j in time) or a chemical process that only slowly generates the particles j.

9.4 Birth and Growth of Brownian Clusters

By flocculation or aggregation we refer to Brownian particles that stick together while keeping their identity in the form of their shape, in contrast to droplets which merge together in a coalescence process. When the colloids strongly attract each other such that each Brownian encounter leads to a permanent aggregate, we speak of *fast* flocculation.

Early-phase flocculation. The initial stage of fast flocculation is the irreversible aggregation of two monomeric particles into a dimer. The dimerization kinetics is that of any irreversible 'bi-molecular reaction' between species i and j:

$$\frac{dc_i}{dt} = \frac{dc_j}{dt} = -k_{ij}\, c_i\, c_j \tag{9.24}$$

Here c_i and c_j are bulk number concentrations of species i and j (the subscript ∞ denoting bulk values has been dropped). The rate constant k_{ij} of this second order reaction directly follows from the flux in Eq. (9.16), because the total collision frequency between particles i and j equals $J(j \to i)c_i$. Since every collision removes a free i and j particle we have:

$$\frac{dc_i}{dt} = \frac{dc_j}{dt} = -J(j \to i)c_i, \tag{9.25}$$

which implies for the rate constant:

$$k_{ij} = J(j \to i)/c_j = 4\pi D_{ij} R_{ij}, \tag{9.26}$$

where we have substituted the diffusive flux $J(j \to i)$ from Eq. (9.16). This rate constant, first derived by Smoluchowski, has the remarkable feature that for monodisperse particles it is independent of particle size, for if we substitute $R_{ij} = 2R_1$ and $D_{ij} = 2D_1$ it turns out that:

$$k_{11} = \frac{8kT}{3\eta} \tag{9.27}$$

This size independence suggests that k_{11} should also give a reasonable estimate for diffusion-controlled reactions between small molecules or ions. For example, for

$$OH^- + NH_4^+ \xrightarrow{k_r} NH_3 + H_2O,$$

the rate constant is $k_r = 5.6 \times 10^{-17}$ m^3/s. From (9.27) we obtain for water ($\eta = 0.84$ mPa s) at 298 K: $k_r = 1.2 \times 10^{-17}$ m^3/s, which is indeed correct in order of magnitude.

Non-spheres. In addition to its weak size independence, the rate constant in (9.26), derived for spheres, is also fairly insensitive to the shape of Brownian particles. We will illustrate this for the (very non-spherical) case of thin rods with diameter D and length $L \gg D$. The orientationally averaged diffusion coefficient of a thin rod is[1]:

---

[1] J. K. G. Dhont, *An Introduction to the Dynamics of Colloids* (Elsevier, Amsterdam, 1996).

$$D = \frac{kT}{3\pi \eta L} \ln\left(\frac{L}{D}\right), \quad \text{for } \frac{L}{D} \gg 1 \qquad (9.28)$$

A freely rotating rod sweeps by rotational diffusion a spherical volume with diameter L. Brownian encounters between two rods may occur when their 'rotation volumes' overlap. Thus the collision distance for the rods is about $R_{11} \sim L$ which on substitution together with (9.28) in (9.26) yields:

$$k_{11} = 8\pi D R_{11} = \frac{8kT}{3\eta} \ln\left(\frac{L}{D}\right) \qquad (9.29)$$

So even for thin Brownian rods, (9.27) provides a reasonable estimate of the rate constant; the rod's aspect ratio only slightly enhances k_{11} via its logarithm. The implication is that kinetic results derived in what follows for spheres, also approximately hold for non-spheres. The underlying reason is that non-sphericity increases the average collision distance R_{ij} between colloids but its enhancing effect on collision frequency is compensated by the slower diffusion of non-spheres.

Singlets half-time. Even though the rate constant is independent of particle size, the particle number density due to flocculation decreases in time at a rate that strongly depends on the sphere radius R, as can be seen as follows. Consider the initial stage of flocculation in which only doublets of spheres are formed. Equation reads for identical spheres:

$$\frac{dc_1}{dt} = -k_{11} (c_1)^2 \qquad (9.30)$$

The solution of (9.30) shows that in the initial stage the concentration $c(t)$ of free singlet spheres decreases as

$$c(t) = \frac{c_0}{1 + (t/t_{1/2})} \qquad (9.31)$$

Here $t_{1/2}$ is the half-life of the singlet spheres and c_0 the singlet number density at $t=0$. For a starting volume fraction $\phi_0 = c_0 (4/3)\pi R^3$ the half-life equals:

$$t_{1/2} = \frac{1}{k_{11} c_0} = \frac{\pi \tau_{CR}}{2\phi_0}; \quad \tau_{CR} = \frac{\eta R^3}{kT} \qquad (9.32)$$

Here τ_{CR} is the configurational relaxation time from Eq. (4.11); note that $t_{1/2}$ approximately equals the Brownian collision time $\tau_{BC} = \tau_{CR}/\phi_0$, that we met earlier in Sect. 4.3. The half-life also turns out to be the reciprocal of the flux in (9.20). It is evident that for a given volume fraction, colloids in the micron size range flocculate relatively slowly. For nano-particles at a volume fraction of say $\phi = 0.01$, rapid flocculation occurs within a split of a second.

Late-stage flocculation. Beyond the initial stage of dimer formation further aggregation of particles by Brownian motion produces triplets, quadruplets etc.

which, in turn, also collide by diffusion to form large clusters. Smoluchowski showed that this—at first sight hopelessly complicated—kinetic problem can be approximately solved as follows. Consider the concentration c_α of aggregates containing α spheres. Such α-mers are formed by the encounters of smaller aggregates, and disappear by the uptake of any other particle or aggregate. The change of α-mer concentration in time is therefore:

$$\frac{dc_\alpha}{dt} = \frac{1}{2} \sum_{i=1}^{\alpha-1} k_{i,\alpha-i} c_i c_{\alpha-i} - \sum_{i=1}^{\infty} k_{i\alpha} c_i c_\alpha \tag{9.33}$$

This equation can be solved easily if we neglect any difference between reaction rate constants k_{ij} and consequently substitute $k_{ij} = k_{11}$:

$$\frac{dc_\alpha}{dt} = \frac{1}{2} k_{11} \sum_{i=1}^{\alpha-1} c_i c_{\alpha-i} - k_{11} c_\alpha \sum_{i=1}^{\infty} c_i \tag{9.34}$$

In terms of the total number density

$$c_{tot} = \sum_{i=1}^{\infty} c_i, \tag{9.35}$$

Equation (9.34) can be written as:

$$\frac{dc_{tot}}{dt} = -\frac{1}{2} k_{11} c_{tot}^2, \tag{9.36}$$

with the solution

$$c(t) = \frac{c_0}{1 + t/t_{1/2}} \tag{9.37}$$

Here the half-life equals

$$t_{1/2} = \frac{\pi \tau_{CR}}{\phi_0}, \tag{9.38}$$

which is twice the half-life of singlet spheres in Eq. (9.32). Apart from the total particle number density, we can also evaluate the concentration all the various particle species (α − mers) in time. From Eq. (9.34):

$$\frac{dc_1}{dt} = -k_{11} \sum_{i=1}^{\infty} c_i c_1; \quad \frac{dc_2}{dt} = \frac{1}{2} k_{11} c_1^2 - k_{11} \sum_{i=1}^{\infty} c_i c_2, \text{ etc.} \tag{9.39}$$

Again all rate constants are equal: $k_{ij} = k_{11}$. This leads to the concentrations of the various species in Fig. 9.3. The assumption that all rate constants equal k_{11} seems

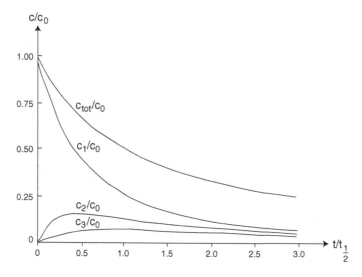

Fig. 9.3 Change in species concentration in time according to Eq. (9.39)

drastic. It implies, for example, that the rate constant for *irregular* aggregates of, say, 10 particles equals that of one single sphere. Note, however, that the diffusion coefficient D of a cluster is inversely proportional to the typical cluster size R_c. This implies that the rate constant $k \sim DR_c$ is indeed fairly insensitive to the shape and size of the aggregates that form in the flocculation process.

Exercises

9.1 Derive the equivalent of (9.16) for Brownian motion on a flat plane, for discs with radius R_j towards a target disc with radius R_i. Consider the limit $\delta \to \infty$. Conclusion?

9.2 Show that (9.17) is indeed correct for spheres that diffuse independently from each other.

9.3 Show that (9.20) is the minimum of (9.19).

9.4 Calculate the half-life for the flocculation of identical spheres in the initial stage, for $\phi = 0.01$ and $R = 10$ nm, respectively, $R = 10\,\mu$m.

9.5 Consider a mixture of spheres with a certain distribution in the sphere radius. Show that by diffusional growth the distribution will always sharpen.

9.6 Calculate the capillary radius a_B for which the time ratio in (9.5) equals unity for the colloidal C-sphere ($R = 100$ nm) and the nano-sphere ($R = 5$ nm) from Table 1.1.

9.7 An adult in rest has a blood pressure of about 0.13 bar. Which water height h in Fig. 9.1 produces the same pressure?

9.8 The following numbers were obtained by counting free silica spheres in water ($T = 298$ K, viscosity 0.89 mPa s) during flocculation by excess sodium chloride:

$$t/\text{min} \quad \quad 0 \quad 2 \quad 4 \quad 7 \quad 12 \quad 20$$
$$c/10^8\text{cm}^{-3} \quad 100 \quad 14 \quad 8.2 \quad 4.6 \quad 2.8 \quad 1.7$$

Calculate the second-order rate constant k_{11} and compare it with the prediction based on the assumption that the flocculation is a diffusion-controlled process.

9.9 Colloidal spheres with radius $R = 500$ nm and volume fraction of $\phi_1 = 0.01$ are mixed in solution with small nano-particles with radius $r = 10$ nm and volume fraction $\phi_2 = 0.05$ Calculate the Brownian encounter frequency between small spheres and one big sphere.

References

For the original work of Smoluchowski on his diffusion model for coagulation see M. von Smoluchowski, *Drei Vorträge über Diffusion, Brownsche Molekularbewegung und Koagulation von Kolloidteilchen*, Phys. Z. **17** (1916) 557–599, and M. von Smoluchowski, *Versuch einer mathematischen Theorie der Koagulationskinetik kolloider Lösungen*, Z. Phys. Chem. **92** (1917) 129–168.

Brownian encounters in biology are examined in H. C. Berg, *Random Walks in Biology* (Princeton University Press, expanded edition, 1993).

Section 9.2 treats diffusion towards a homogeneous sphere; for Brownian motion towards a patchy cell with receptors on its surface see H. C. Berg and E. M. Purcell, *Physics of Chemoreception*, Biophysical Journal **20** (1977) 193–219.

For a review on precipitation, nucleation and growth of colloids see: A. P. Philipse, *Particulate Colloids: Aspects of Preparation and Characterization*, in: J. Lyklema (Ed.), *Fundamentals of Colloid and Interface Science*, Vol. IV (Elsevier, 2005).

Chapter 10
Random Walks in External Fields

Brownian motion comprises a sequence of randomly oriented steps (Fig. 4.5)—which is why it is often compared to the walk of a (very) drunk man who stumbles in every direction with equal probability. The man only makes these unbiassed steps when stumbling on a wide, horizontal street; on a sloping street his down-hill steps are more probable than up-hill ones. And when the man has accidentally entered a narrow alley, movements to the left and to the right are confined by hard walls that impose on our pitiable friend a (quasi) one-dimensional random walk parallel to these walls.

Slopes and walls are examples of external fields that break the overall, on average radial symmetry of Brownian displacements in absence of external forces. Associated with such a field is a potential; gravitational potential energy in the case of a sloping street, and a steep repulsive 'hard wall' potential experienced by the drunkard when he (of course, at random moments in time) attempts to penetrate a solid wall.

In this Chapter we will study colloids that diffuse up- or downhill a gradient in a potential V; this gradient corresponds to an external force K on each colloid:

$$\vec{K} = -\vec{\nabla} V; \quad \oint dV = 0 \tag{10.1}$$

Force K is a conservative one, implying that the integral of V over a closed path is zero, as indicated, because the potential is a variable of state. In addition to gravitational and hard-wall potentials, colloids may also be subjected to electrical or magnetic forces, the latter being examined in Sect. 10.4 on Brownian rotations in a magnetic field. However, an external force on a colloid can also stem from another colloid, as in the case of charged spheres to be addressed in Sect. 10.2.

© Springer Nature Switzerland AG 2018
A. P. Philipse, *Brownian Motion*, Undergraduate Lecture Notes in Physics,
https://doi.org/10.1007/978-3-319-98053-9_10

10.1 One-Dimensional Diffusion

In Fig. 10.1 the Brownian motion of particles leads to a net diffusive transport in the x-direction, from a source with constant concentration ρ_0 at $x = a$ to a sink with concentration $\rho = 0$ at $x = a + L$. In absence of a potential, the diffusion flux and the concentration gradient in the steady state are:

$$j_d = -D\frac{d\rho(x)}{dx} = D\frac{\rho_0}{L}; \quad \rho(x) = \rho_0\left(1 - \frac{x}{L}\right) \tag{10.2}$$

The external force $K = -dV(x)dx$ on each particle produces the convective flux

$$j_c = -\rho(x)u = -\frac{\rho(x)}{f}\frac{dV(x)}{dx} \tag{10.3}$$

Thus the total steady-state flux $j = j_d + j_c$ is given by

$$j = -D\left[\frac{d\rho(x)}{dx} + \frac{\rho(x)}{kT}\frac{dV(x)}{dx}\right] = \text{constant} \tag{10.4}$$

In thermodynamic equilibrium the total flux is zero; for $j = 0$ the solution of (10.4) is the equilibrium number density profile:

$$\rho_{eq}(x) = \rho_0\exp[-V(x)/kT], \tag{10.5}$$

which is an instance of the Boltzmann distribution. To find the concentration profile $\rho(x)$ in the non-equilibrium steady-state $(j \neq 0)$ we note that the profile must reduce to (10.5) when the flux is zero; so as a trial solution we choose:

Fig. 10.1 Particles diffuse in the x-direction from a source with constant concentration ρ_0 across a potential barrier of width L into a sink where the particle concentration remains zero

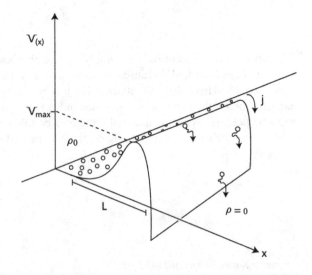

$$\rho(x) = \gamma(x)\rho_{ev}(x) = \gamma(x)\exp[-V(x)/kT], \tag{10.6}$$

where $\gamma(x)$ is a function that in equilibrium equals the concentration ρ_0. Substitution of (10.6) in (10.4) leads to the following differential equation for $\gamma(x)$:

$$\exp[-V(x)/kT]\frac{d\gamma(x)}{dx} = -\frac{j}{D} \tag{10.7}$$

In view of the boundary condition $\gamma(x = a) = \rho_0$ Eq. (10.7) yields:

$$\gamma(x) = \rho_0 - \frac{j}{D}\int_a^x \exp[V(x')/kT]dx' \tag{10.8}$$

The magnitude of the steady-state flux follows from the second boundary condition, namely that $\gamma(x = a + L) = 0$:

$$j = \frac{D\rho_0}{\int_a^{a+L}\exp[V(x)/kT]dx} \tag{10.9}$$

The steady-state concentration profile is obtained by substitution of (10.8) and (10.9) in (10.6):

$$\frac{\rho(x)}{\rho_{eq}(x)} = 1 - \frac{\int_a^x \exp[V(x')/kT]dx'}{\int_a^{a+L}\exp[V(x)/kT]dx} \tag{10.10}$$

Here $\rho_{eq}(x)$ is the equilibrium Boltzmann distribution from (10.5). Note that for $V(x)=0$ we recover the flux and the linear concentration profile in Eq. (10.2).

The delay factor. With respect to kinetics, the essential point is that the effect of a potential is equivalent to a rescaling of the diffusion coefficient. We can rewrite (10.9) as:

$$j = D_{eff}\frac{\rho_0}{L}; \quad D_{eff} = \Theta_1 D \tag{10.11}$$

Here D_{eff} is an effective diffusion coefficient that accounts for the retardation of particle transport due to the external potential. In (10.11) the upper case Greek letter Θ denotes the dimension-less delay factor

$$\Theta_1 = \frac{1}{L}\int_a^{a+L}\exp[V(x)/kT]dx \tag{10.12}$$

To make an estimate of this factor we note that for a high repulsive barrier V_{max} (see Fig. 10.1), the integral in (10.12) approximately equals $(L/2)\exp(V_{max}/kT)$. Thus:

$$D_{\text{eff}} \approx \frac{1}{2} D \, \exp[-V_{\text{max}}/kT], \quad \text{for } V_{\text{max}} \gg kT \tag{10.13}$$

This result reminds of the Arrhenius equation in chemical reaction kinetics, where reaction rates are exponentially retarded by an activation energy barrier. Equation (10.13) informs us that, independent of the detailed shape of the potential $V(x)$, a repulsive barrier in the range 5–10 kT suffices to practically eliminate Brownian diffusion across the barrier in Fig. 10.1.

Age of a Brownian floc. Particle clusters, of course, may also fall apart due to thermal motion provided the attractive well is not too deep. We can estimate the life time of a doublet from the time it takes for a particle to diffuse out of a well with depth V_{max} and a width comparable to its own radius:

$$\tau \sim \frac{R^2}{D_{\text{eff}}} \sim \frac{\eta R^3}{kT} \exp[V_{\text{max}}/kT] = \tau_{\text{CR}} \exp[V_{\text{max}}/kT], \tag{10.14}$$

where τ_{CR} is the configurational relaxation time for the colloids in absence of any force field. This scaling relation also gives an indication for the temporal stability of larger clusters or aggregates or particle gels. Such non-equilibrium structures can be called permanent when the time τ from (10.14) is very much larger than an observation time which is about 10^{-2}–10^{-3} s for dynamic light scattering and minutes for optical microscopy.

10.2 Radial Brownian Motion and Colloidal Stability

When two uncharged colloidal particles meet via Brownian motion, they will stick together by the van der Waals attraction. The kinetics of this irreversible aggregation was the subject of Sect. 9.2. However, when colloids carry electric surface charge, they are surrounded by a diffuse clouds of oppositely charged counter ions that upon overlap generate an repulsion[1]. When this repulsion outweighs the van der Waals attraction the colloids are said to be stable, see also Fig. 10.2. This stability is not a thermodynamic one but, instead, a kinetic one: for the large majority of Brownian encounters the kinetic energy of the colloids is insufficient to approach each other closely enough to feel the pull of the van der Waals force. Hence, after the encounter colloids just freely diffuse away.

The radial delay factor. To calculate the factor by which a repulsion between colloidal spheres slows down Brownian collision frequencies, we consider a sphere diffusing in a radial potential $V(r)$ induced by a sphere that is centered at the origin. The steady state for radial diffusion of spheres towards the origin is

---

[1] Entropic because the repulsion stems from compression of an ideal ion gas between two approaching surfaces.

Fig. 10.2 Due to a potential barrier V_{max} two spheres, diffusing in three dimensions, cannot approach each other sufficiently close to irreversibly form a dimer by the van der Waals attraction

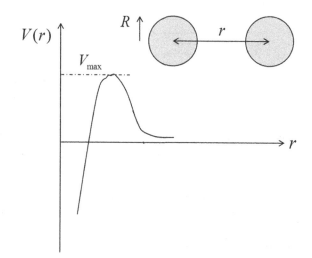

$$J = 4\pi r^2 D_{12}\left[\frac{d\rho(r)}{dr} + \frac{\rho(r)}{kT}\frac{dV_{12}(r)}{dr}\right] \qquad (10.15)$$

Here J is the total particle flux per second through a spherical envelope with radius r, $V_{12}(r)$ is the interaction potential between spheres 1 and 2, and D_{12} is their relative diffusion coefficient. For the boundary conditions $\rho(r=R_{12})=0$, and $\rho=0$ in the bulk at infinite r, solution of (10.15) (exercise 10.1) leads to the steady-state flux:

$$J = \Theta_r\rho_0 4\pi D_{12}R_{12}; \quad R_{12} = R_1 + R_2, \qquad (10.16)$$

where R_{12} is the center-to-center contact distance. The dimension-less delay factor Θ_r is defined by

$$\frac{1}{\Theta_r} = -R_{12}\int_{\infty}^{R_{12}}\frac{\exp[V_{12}(r)/kT]}{r^2}dr, \qquad (10.17)$$

and is the three-dimensional, radial analogue of the delay factor in Eq. (10.12). Note that $\Theta_r = 1$ for $V_{12}(r)=0$ such that the flux J reduces to the flux found earlier in Chap. 9 for spheres without interaction (except for an irreversible sticking at contact).

Coulomb repulsions. An interaction for which the delay factor can be easily evaluated is the repulsion between spheres 1 and 2 that carry, respectively, z_1 and z_2 elementary charges on their surface. The Coulomb repulsive interaction energy between the two spheres at center-to-center distance r is accordingly:

$$\frac{V_{12}(r)}{kT} = \frac{z_1z_2e^2}{4\pi\varepsilon\varepsilon_0 rkT} = z_1z_2\left(\frac{L_B}{r}\right); \quad L_B = \frac{e^2}{4\pi\varepsilon\varepsilon_0 kT} \qquad (10.18)$$

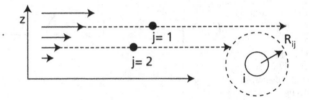

Fig. 10.3 Centers of particles j move in a shear flow towards target particle i. Shear induced flocculation is auto-catalytic: the larger the particle aggregate, the more rapid it will catch other particles in a stirred solution

Here L_B is the Bjerrum length, defined as the distance between two elementary charges e at which their interaction energy equals the thermal energy kT, for charges immersed in a solvent with di-electrical constant $\varepsilon\varepsilon_0$. Solving integral (10.17) for the Coulomb repulsion yields:

$$\Theta_r = \frac{y_{12}}{\exp(y_{12}) - 1}; \quad y_{12} = z_1 z_2 \left(\frac{L_B}{R_{12}}\right) \tag{10.19}$$

For identical spheres with valency z, the delay factor simplifies to:

$$\Theta_r = \frac{y}{\exp(y) - 1}; \quad y = z^2 \left(\frac{L_B}{2R}\right) \tag{10.20}$$

The diffusion flux from (10.16) becomes for identical spheres:

$$J = \rho_0 \frac{8kT}{3\eta}\Theta_r = \rho_0 \frac{8kT}{3\eta}\frac{y}{\exp(y) - 1}; \quad y = z^2 \left(\frac{L_B}{2R}\right) \tag{10.21}$$

This diffusive flux—and hence the collision frequency between spheres—diminishes dramatically with increasing sphere size. For a fixed surface charge density, the sphere valency scales with the sphere radius as $z \sim R^2$ such that $y \sim R^3$ and, consequently, the flux in (10.21) scales as $J \sim R^3 \exp(-R^3)$.

10.3   Brownian Motion in a Shear Flow

In Chap. 9 we have analyzed the fast flocculation of colloids that perform Brownian motion in a quiescent solution. If the solution is stirred, shear forces increase the flocculation rate, because velocity gradients in the solvent increase the collision frequency of the colloids. Suppose a constant velocity gradient $\dot{\gamma} = dv/dz$ is present in the z-direction (see Fig. 10.3). Then the flux of particles j at height z in the direction of sphere i centered at $z = 0$ is:

Fig. 10.4 Sketch accompanying Eq. (10.22)

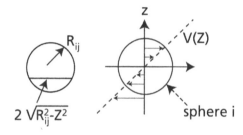

$$c_j v(z) 2\sqrt{\left(R_{ij}^2 - z^2\right)}\,dz; \quad v(z) = \dot{\gamma}\,z \tag{10.22}$$

Integration of this expression (see also Fig. 10.4) gives the total flux of j-particles to the i-sphere:

$$J(j \to i) = 4c_j \dot{\gamma} \int_0^{R_{ij}} z\sqrt{\left(R_{ij}^2 - z^2\right)}\,dz = \frac{4}{3}\dot{\gamma} R_{ij}^3 c_j \tag{10.23}$$

The corresponding rate constant is:

$$k_{ij} = \frac{4}{3}\dot{\gamma}\,R_{ij}^3 \tag{10.24}$$

We compare this 'shear-induced' rate constant to the purely diffusional rate constant derived in Chap. 9:

$$\frac{(k_{ij})_{\text{shear}}}{(k_{ij})_{\text{diff}}} = \frac{\dot{\gamma} R_{ij}^2}{3\pi D_{ij}} \approx \frac{4R^3 \eta \dot{\gamma}}{kT} = 4\tau_{\text{CR}}\dot{\gamma} \tag{10.25}$$

Clearly for large particles ($R > 1\,\mu$m) shear-induced flocculation becomes important. In practice, this type of flocculation appears to have an auto-catalytic character: once flocculation has started (either by diffusion or stirring) further stirring strongly accelerates the process. This is because of the R^3—dependence in Eq. (10.25): the larger the aggregate or floc the more rapid it will catch other aggregates in the stirred suspension.

10.4 Brownian Magnets in a Magnetic Field

So far we have considered the effect of an external field on translational particle diffusion; we will now address an example of a field that only couples to rotational Brownian motion. It is a magnetic field that interacts with the magnetic moments of

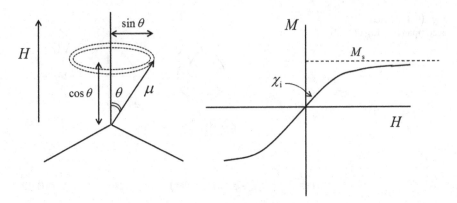

Fig. 10.5 Left: a dipole moment μ has its lowest energy when parallel to the external field H but due to rotational Brownian motions, dipole moments adopt an equilibrium distribution in the polar angular range $\theta = 0$ to $\theta = \pi$. Right: Sketch of a magnetization curve; M is the average magnetization per volume in the direction of $\vec{H}$. Initially M increases linearly with a slope χ_i; upon further increase of H the angular dipole distribution progressively narrows until saturation magnetization M_s is reached

magnetic colloids, for example, colloids composed of magnetic iron-oxide[2]. These colloids have an embedded *permanent* magnetic moment μ and the Brownian motion of such colloids can be envisioned as the thermal tumbling and turning of colloidal compass needles.

Since these needles have no preferred orientations, the net magnetization of a dispersion of magnetic colloids is zero. To generate magnetization, magnetic moments must be forced to line up by means of an external magnetic field $\vec{H}$. When this field is homogeneous, only a torque is exerted on the moments so only rotational diffusion is affected; when the field is inhomogeneous a translational force is exerted on the colloids as well.

Angular distribution function. The energy E of a (point) dipole $\vec{\mu}$ in applied field $\vec{H}$ is given by the vector product

$$E = -\vec{\mu} \cdot \vec{H} \tag{10.26}$$

So the configuration with lowest energy is that of a dipole moment parallel to the external field (Fig. 10.5); this is the ground state for a dipole at zero Kelvin. At finite temperature, however, the dipole moment executes rotational Brownian motion as a result of which dipole orientations adopt an equilibrium distribution of polar angles in the range from $\theta = 0$ to $\theta = \pi$ (Fig. 10.5). Since the ends of dipole vectors having the same orientation θ, end up in an annular area $2\pi\theta d\theta$ (Fig. 10.5), the number of these vectors is proportional to that area. Therefore the distribution function for dipole orientations is of the form:

---

[2]B. Luigjes et al., J. Phys.: Condens. Matter **24** (2012) 245104.

$$P(\theta) = C \, \exp[(\vec{\mu} \cdot \vec{H})/kT]2\pi \, \sin\theta d\theta = C \, \exp[(\alpha \, \cos\theta]2\pi \, \sin\theta d\theta \quad (10.27)$$

Here C is a normalization constant (Exercise 10.5) and α is the ratio of the maximal energy of a dipole in the external field, to the thermal energy kT:

$$\alpha = \frac{\mu H}{kT} \quad (10.28)$$

The Langevin function. The contribution of a dipole to the net magnetization induced by the field H is the component of the dipole moment parallel to the field which, as can be seen in Fig. 10.5, equals $\mu\cos\theta$. The average value of this component is[3] $\mu<\cos\theta>$, where the average cosine of the polar angle between dipoles and field equals (exercise 10.3):

$$< \cos\theta > = \frac{\int_0^\pi \cos\theta \, \exp(\alpha \, \cos\theta) \, \sin\theta d\theta}{\int_0^\pi \exp(\alpha \, \cos\theta) \, \sin\theta d\theta} = \Lambda(\alpha) \quad (10.29)$$

Here the Greek uppercase letter Λ denotes the so-called *Langevin function* which is defined by[4]:

$$\Lambda(\alpha) = \coth(\alpha) - \frac{1}{\alpha} \quad (10.30)$$

For a liquid dispersion with a number density ρ of colloids with an embedded permanent magnetic moment, the net magnetization M per volume of dispersion is:

$$M = \rho\mu \, \Lambda(\alpha) \quad (10.31)$$

At strong external fields such that $\alpha \gg 1$, the Langevin function approaches $\Lambda(\alpha) = 1$, signifying that all dipoles are aligned parallel to the field. Then the colloidal dispersion has reached its maximal or *saturation* magnetization M_s:

$$M_s = \rho\mu, \quad \text{for } \alpha \gg 1 \quad (10.32)$$

In a weak external field ($\alpha \ll 1$) thermal angular displacements of dipoles are significant. From the weak-field limit of the Langevin function (exercise 10.4)

$$\Lambda(\alpha) = \frac{\alpha}{3}, \quad \text{for } \alpha \ll 1, \quad (10.33)$$

it follows that the initial linear slope of the magnetization curve (Fig. 10.5), also referred to as the initial *susceptibility*, is given by:

---

[3] It is assumed here that all dipole moments have the same value μ.
[4] $\coth(x) = \cosh(x)/\sinh(x) = (e^x + e^{-x})/(e^x - e^{-x})$.

$$\chi_i = \frac{M}{H} = \frac{\rho\mu^2}{3kT} \qquad (10.34)$$

Colloid number densities ρ are usually not accurately known[5] but ρ can be eliminated by scaling the measured initial susceptibility on the saturation magnetization:

$$\frac{\chi_i}{M} = \frac{\mu}{3kT} \qquad (10.35)$$

This determination of magnetic moments only works if the distribution of magnetic particle moments is sufficiently narrow. In a mixture of magnetic moments, the large dipole moments will align at low fields and, hence, dominate χ_i whereas weak moments will only start to contribute on approach of the saturation magnetization.

10.5   Gravity

In practice, many colloids (like those in paints, clays and dairy products) have enough mass to feel the pull of the Earth. Hence the haphazard Brownian motion of these colloids competes with a particle flux that is directed towards the center of Earth. In equilibrium we have the force balance

$$\frac{d\pi(h)}{dh} = -(\Delta m)g\rho(h) \qquad (10.36)$$

Here $\pi(h)$ and $\rho(h)$ are, respectively, osmotic pressure and particle number density at altitude h, Δm is the buoyant particle mass and g is the gravitational acceleration.

Equation of state. The force balance (10.36) harbors the osmotic equation of state (OES), *i.e.* the dependence of osmotic pressure $\pi(\rho)$ on colloid concentration. Integration of (10.36) yields:

$$\pi(h) = (\Delta m)g \int_h^\infty \rho(h')dh' \qquad (10.37)$$

In words: the osmotic pressure exerted by particle number density $\rho(h)$ at height h carries the buoyant weight of particles in the column from h to infinity[6]. From an experimentally determined concentration profile $\rho(h)$ one can then retrieve the OES[7]. So here the gravitational field is employed to gauge the pressure exerted by

---

[5]The experimental concentration measure is the colloid *weight* concentration; its conversion to a number density requires the particle mass density which is not easy to determine.

[6]'Infinity' is here an altitude high enough to leave the particle profile behind and enter pure solvent.

[7]For an example of this osmotic pressure measurement by centrifugation, see Piazza R. Bellini T. and Degiorgio V. *Equilibrium Sedimentation Profiles of Screened Charged Colloids. A Test of the Hard-Sphere Equation of State*, Phys. Rev. Lett. **71** (1993) 4267.

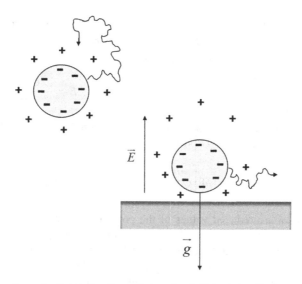

Fig. 10.6 For a charged colloid in free Brownian motion (left) its thermally fluctuating counter ion cloud remains on average spherically symmetric. The gravity field (right) biases diffusive steps of colloids towards the earth but has no effect on the virtually weightless ions that tend to stray into free space. Electro-neutrality, however, orders the ions to stay in the colloid's vicinity. In equilibrium, centers of positive and negative ion distributions no longer coincide, leading to an electric field that lifts up the colloid

Brownian particles; in Sect. 11.3 we will witness the use of an external field in the form of a membrane for that purpose.

Barometric profile. We continue with non-interacting colloids, with the EOS $\pi = \rho kT$, for which the solution of (10.36) is the barometric profile that we already encountered in Sect. 2.3:

$$\rho(h) = \rho_0 \exp\left(\frac{-h}{l_g}\right); \quad l_g = \frac{kT}{(\Delta m)g}, \tag{10.38}$$

where ρ_0 is the colloid number density at zero altitude. One interesting—and often overlooked—point is that ideality of colloids is a necessary condition for the barometric profile (10.38) to hold, but it is not a sufficient one: the colloids must also be uncharged. When colloids carry electrical charge the equilibrium sedimentation-diffusion (SD) profile is non-barometric, even at infinite dilution. The physical reason for this is that gravity induces in the SD-profile an internal electric field that lifts up the Brownian particles. This interesting phenomenon is due to the huge mass difference between colloids and ions, as explained in the legend of Fig. 10.6.

Charged colloids. Consider a solution with a mixture of three species: negatively charged colloids with charge number z and number density ρ, monovalent cations with concentration c_+ and monovalent anions with concentration c_-. The solution is in equilibrium with an external reservoir with total ion concentration $2c_s$—in a sedimentation cell this reservoir is the supernatant solution devoid of colloids at high altitude. Suppose an electrical field with magnitude E acts on the colloids than the force balance on non-interacting colloids reads:

$$kT\frac{d\rho(h)}{dh} = -(\Delta m)g\rho(h) - zeE\rho(h); \quad E = -\frac{d\Psi}{dh} \tag{10.39}$$

Here e is the proton charge and ψ the electrical potential and E the electric field. The solution of (10.39) is the SD-profile:

$$\rho(h) = \rho_0 \exp\left[\frac{-h}{l_g} + z(\varphi - \varphi_0)\right]; \quad \varphi = \frac{e\Psi}{kT} \tag{10.40}$$

Here φ and φ_0 are reduced values of the potentials ψ and ψ_0 at, respectively, altitudes h and $h=0$. Further, ρ_0 is the colloid number density at $h=0$. The profile in (10.40) is non-exponential because the electrical potential depends on the colloid concentration and, consequently, also on altitude. The potential follows from the force balance for ideal, weightless ions in an electric field:

$$-kT\frac{dc_{\pm}}{dx} \pm ec_{\pm}E = 0, \tag{10.41}$$

which entails the equilibrium ion density profiles:

$$c_{\pm} = c_s\exp[\mp\varphi] \tag{10.42}$$

The SD-profile as a whole must be electrically neutral which implies:

$$-z\rho(h) + c_+ - c_- = 0 \tag{10.43}$$

From (10.42) and (10.43) we obtain the potential as:

$$\varphi = \mathrm{arcsinh}\left(-\frac{z\rho(h)}{2c_s}\right) \tag{10.44}$$

Equations (10.40) and (10.44) describe a concentration profile that is inflated in comparison to the barometric profile in (10.38). This is because the macroscopic electric field inside the profile exerts a force on the colloids that is oppositely directed to gravity (Fig. 10.6). Thus the field reduces the buoyant mass of the particles. This reduction becomes more significant at lower salt concentration and vanishes at high ionic strength (exercise 10.6). If the induced field compensates the particle mass (exercise 10.6) the colloids are weightless and freely perform Brownian motion as if no external field was present.

10.6 Exercises

10.1 Solve the differential equation in (10.15) $V(r)$ to find the steady-state flux in (10.16). Employ the boundary conditions $\rho(r=R_{12})=0$, and $\rho=0$ in the bulk at infinite r.

10.2 Derive the delay factor in (10.17) for ions with equal radii R and valency z. Also verify (10.19).

10.3 Verify the calculation of $<\cos\theta>$ in (10.29).

10.4 Verify the weak-field limit of the Langevin function in (10.33).

10.5 Determine the normalization constant C in (10.27).

10.6 (a) Show that for high salt concentration c_s the SD-profile in (10.40) approaches the barometric distribution (10.38). (b) Estimate the field strength needed to reduce the buoyancy in water of C-spheres from Table 1.1 to zero. Assume that the spheres carry one negative surface charge per nm^2.

10.7 (a) The barometric profile is a Boltzmann distribution, so the distribution function has the form $P(h)=C\,\exp(-Energy/kT)$; $C=$ constant. Here $P(h)$ is the distribution function for particles at height h; k is the Boltzmann constant. Which energy term should we substitute here for ideal, uncharged particles?
(b) What is the meaning of $P(h)\,dh$?
(c) Calculate the constant C via the normalization: $\int P(h)dh=1$
(d) Calculate the average height $<h>$ of the particles above the Earth's surface located at $h=0$.
(e) Evaluate $<h>$ for nitrogen particles and for colloids in the form of small water droplets with a diameter of 1 μm and mass density 1 g/ml in an atmosphere at $T=298$ K. Discuss the two outcomes.
(f) Derive a formula for the root-mean-square height of particles in a barometric distribution.
(g) Compute the average gravitational potential energy of particles in the distribution.

References

Brownian motion in a force field is treated in: P. Debye, *Molecular Forces*, Wiley 1967; N. Fuchs, *Über die Stabiliät und Auflading der Aerosole*, Z. Phys. **89** (1934) 736–743, and H. A. Kramers, *Brownian motion in a field of force and the diffusion model of chemical reactions*, Physica VII (4) (1940) 284–304.

The classic text on the theory of colloidal stability: E. J. Verwey and J. Th. G. Overbeek, *Theory of the Stability of Lyophobic Colloids* (Elsevier, 1948). Reprinted Dover Publications, Inc. 1999.

A monograph on ferro-fluids (concentrated dispersions of magnetic iron-oxide colloids) is: R. E. Rosensweig, *Ferrohydrodynamics* (Cambridge University Press, 1985). Reprinted Dover Publications, Inc. 1997.

For the electric field in the sedimentation profile of charged colloids, see: R van Roij, *Defying entropy with gravity and electrostatics*, J. Phys.: Condensed Matter **15** (2003) S3569; M. Rasa and A. P. Philipse, *Evidence for a macroscopic electric field in the sedimentation profiles of charged macromolecules*, Nature, 2004, **429**, 857.

Chapter 11
Brownian Particles and Van 't Hoff's Law

The inherent thermal motion of Brownian particles brings about both diffusion and osmotic pressure, phenomena that for non-interacting particles are quantified by, respectively, Einstein's diffusion coefficient and Van 't Hoff's law:

$$D = \mu kT; \quad \pi = \rho kT \tag{11.1}$$

The diffusion coefficient of one colloid is found by multiplying the thermal energy kT with the mobility coefficient[1] μ, determined by the surrounding medium; for the osmotic pressure π jointly exert by the colloids, the multiplier is the colloid number density ρ. So one can say that diffusion and pressure of ideal particles are two sides of the same thermal coin. This is also illustrated by Einstein's derivation of the diffusion coefficient $D = \mu kT$ in Chap. 6: a crucial step in the derivation is the assumption of the validity of Van 't Hoff's osmotic pressure law $\pi = \rho kT$, via which the thermal energy kT enters the diffusion coefficient. One could also reverse the argument and assume that $D = \mu kT$: then analysis of the equilibrium profiles from Fig. 6.1 yields Van 't Hoff's law.

There is something to say to proceed from π to D rather than *vice versa* because we have independent derivations of Van 't Hoff's osmotic pressure law. Chapter 3, for example, showed how the ideal pressure law for gases and solutions can be found from the momentum transport by thermal particles. We will discuss in this Chapter two additional derivations of Van 't Hoff's law that are quite different in nature from the kinetic treatment in Chap. 3. The first is based on bulk thermodynamics for the solvent in an osmotic equilibrium; the second is a statistical argument that explicitly invokes the Brownian particles to calculate the pressure they exert. For the 'solvent route' some results from the thermodynamics of dilute dilutions are needed that will be reviewed first.

---

[1] Note that here μ denotes the mobility for both translational and rotational motions.

© Springer Nature Switzerland AG 2018
A. P. Philipse, *Brownian Motion*, Undergraduate Lecture Notes in Physics,
https://doi.org/10.1007/978-3-319-98053-9_11

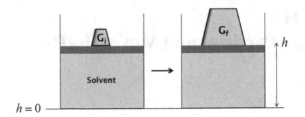

Fig. 11.1 Work done to increase the pressure on an incompressible solvent is equivalent to the work needed to lift extra weight $G_f - G_i$ to height h

11.1 Thermodynamics of Dilute Solutions

Osmosis is the spontaneous mixing of solution and solvent and, as any naturally occurring process, is able to deliver work. The maximal work, it will be recalled, is the work produced in a reversible, quasi-static process. In what follows the term 'work' always denotes this reversible work. The other work term involved in osmosis, in addition to mixing, relates to the increase of pressure on a solution.

Pressure-work. We evaluate the work needed to increase the pressure p in an incompressible solution from an initial value p_i to a final state with value p_f. For the path that leads from state i to state f we select the following gravitational route. A weight G_i is brought from height $h = 0$ to height h, on a piston resting on an incompressible fluid, see Fig. 11.1. The required work equals the increase in potential energy of the weight in the gravity field:

$$w_i = G_i h = p_i A h = p_i V, \tag{11.2}$$

where A is the piston area and V the solution volume. Next the final state is established by lifting a weight G_f to height h which involves the work:

$$w_f = G_f h = p_f V \tag{11.3}$$

Hence the reversible work needed to go from initial to final state is

$$w_p = w_f - w_i = (p_f - p_i) n \overline{V}, \tag{11.4}$$

for n moles of incompressible solvent with molar volume $\overline{V}$. Note that this pressure-work is positive: work has to be performed on the solvent which by convention is counted as positive.

Dilution-work. The reversible work w_d associated with spontaneous dilution of a solution by solvent can be evaluated employing the thermodynamic cycle depicted in Fig. 11.3. Here one mole of solvent is evaporated from a solution to its equilibrium vapor phase (step E), then transferred to the vapor phase above pure solvent (step I) and condensed onto that solvent (step C). The cycle is closed by relocating the

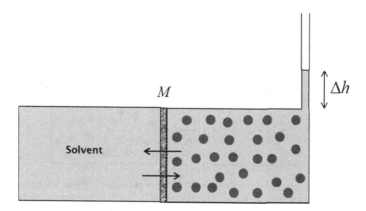

Fig. 11.2 The hydrostatic pressure exerted by the solvent column at the right equals the osmotic pressure of the solution, and counteracts the spontaneous migration of solvent from left to right. In osmotic equilibrium the net water flux across the membrane is zero

solvent to the solution (step D). Since E and C are opposite steps taken in a liquid-vapor equilibrium they involve no net work. Since the total work of the closed cycle must be zero, work w_d in step D is the opposite of the work in step I. The latter follows from Raoult's law for the vapor pressure of a solution.

Raoult's law. Consider (Fig. 11.3) the equilibrium between a solution with solvent mole fraction x_s and its vapor with pressure p. The molecules in the vapor, and the solutes in solution are ideal particles. For water as solvent, equilibrium between solution (L) and the vapor (G) phase can be represented by:

$$H_2O|_L \rightleftarrows H_2O|_G, \tag{11.5}$$

with an equilibrium constant

$$K(T) = \frac{[H_2O]_G}{[H_2O]_L} \propto \frac{p}{[H_2O]_L}, \tag{11.6}$$

that, as any equilibrium constant, depends on temperature only. In a mixture of n_s water molecules and n_f solute particles, the evaporation of water will be a fraction $n_s/(n_s + n_f)$ of that from pure water. For $K(T)$ in (11.6) to remain constant also the vapor pressure in (11.6) has to lower by the same fraction. Thus if p^* is the vapor pressure of pure water, the vapor pressure of the solution at the same temperature is:

$$p = \frac{n_s}{n_s + n_f}p^* = x_s p^*, \tag{11.7}$$

where x_s is the solvent mole fraction in the solution. Equation (11.7) is known as Raoult's law for the vapor pressure of a solution—which is always below the solvent

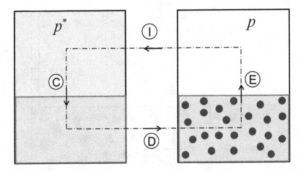

Fig. 11.3 Left: a pure solvent in equilibrium with its vapor pressure p^*. Right: a solution generates a lower equilibrium vapor pressure $p < p^*$. The work done in step D to transfer one mole of solvent to a solution is the opposite of the work done to increase in step I in which the pressure of one mole of vapor is raised from p to p^*

vapor pressure. Employing Raoult's law we can infer that the work needed to bring n moles of ideal gas from pressure p to p^* (step I in Fig. 11.3) is:

$$w = - \int p\mathrm{d}V = nRT \int_{p}^{p^*} p^{-1}\mathrm{d}p = -nRT \ \ln x_s \qquad (11.8)$$

The work of dilution (step D in Fig. 11.3) therefore equals:

$$w_\mathrm{d} = nRT \ \ln x_s \qquad (11.9)$$

The dilution work is negative, indicative of the spontaneous mixing of solvent and solution that can deliver work to the surroundings.

11.2   Osmotic Pressure Gauged via the Solvent

The spontaneous migration (osmosis) of solvent into a concentrated solution builds up a counter-acting hydrostatic pressure such that in osmotic equilibrium (Fig. 11.2) the net water flux is zero. Osmosis involves two work terms namely the work w_p in (11.4) needed to increase the pressure on a solution and, secondly, the work w_d from Eq. (11.9) associated with the spontaneous dilution of a solution.

Osmotic equilibrium. In equilibrium there is no net transport of solvent across the membrane in Fig. 11.2 so the overall work must be zero:

$$w_\mathrm{p} + w_\mathrm{d} = (p_f - p_i)n\overline{V} + nRT \ \ln x_s = 0 \qquad (11.10)$$

Since the pressure drop across the membrane in Fig. 11.2 is $p_f - p_i = \pi$ it follows from (11.10) that the osmotic pressure exerted by ideal solutes equals:

$$\pi = -\frac{\ln x_s}{V} RT \qquad (11.11)$$

The mole fraction of solutes equals $x_f = 1 - x_s$; since for a dilute solution x_s is a very small fraction, the logarithmic term can be linearized:

$$\pi = -\frac{\ln(1 - x_f)}{V} RT = \frac{x_f}{V} RT, \quad \text{for } x_f \ll 1 \qquad (11.12)$$

Since for a very dilute solution of n_f moles in n_s moles of solvent

$$x_f = \frac{n_f}{n_f + n_s} = \frac{n_f}{n_s}, \quad \text{for } n_f \ll n_s, \qquad (11.13)$$

it follows that

$$\pi = \rho kT, \qquad (11.14)$$

where $\rho = n_f/V$ is the number density of solute particles. We have derived here Van 't Hoff's ideal pressure law (11.14) from the thermo-dynamic equilibrium of the solvent. Though Brownian particles are not explicitly addressed in the derivation above, evidently the solutes in Fig. 11.2 must be in thermal motion. If by some unpleasant act of sorcery, colloids would be deprived of their Brownian motion and reduced to static entities, the pressure difference in Fig. 11.2 would disappear, and the solution would not be able to draw in solvent.

11.3 Osmotic Pressure from Brownian Motion; Vrij's Statistical Approach

In Sect. 11.2 we treated osmotic pressure via a thermodynamic point of view, as was also done by Van 't Hoff himself [2] (Fig. 11.4). Here we will, following A. Vrij — see references, derive Van 't Hoff's law via a statistical approach to Brownian motion of particles in the neighborhood of the same semi-permeable membrane employed for the thermodynamic route in Sect. 11.2. Such a membrane can be seen as a field of force that only affects the Brownian particles, not the solvent in which they are immersed.

Membrane force field. Suppose a membrane with thickness $x = \delta$ has its center plane located at $x = 0$ (Fig. 11.5). Selectivity of the membrane implies that colloids arriving by diffusion at location $x \approx \delta$ will experience a sharply rising repulsion. At distances $x \gg \delta$ any repulsion between membrane and colloid obviously vanishes.

---

[2]J. H. Van 't Hoff, *Zeischrift für physikalische Chemie*, **vol. i.** (1887), pp. 481–508.

Fig. 11.4 Jacobus Henricus Van 't Hoff (1872–1911) received in 1905 the first Chemistry Nobel prize for his work on osmotic pressure and thermo-dynamics of dilute solutions

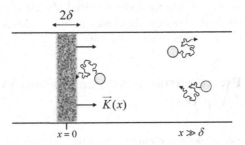

Fig. 11.5 A semi-permeable membrane exerts a force $K(x)$ on colloids that is steeply repulsive for colloids approaching the membrane within a distance $x \approx \delta$, and that vanishes far away from the membrane. These two features of $K(x)$ suffice to calculate the osmotic pressure exerted by the colloids, see Sect. 11.3

Imagine a container of (further unspecified) thermal background particles that exert a pressure p on both sides of a large vertical membrane suspended in the container. We add Brownian particles to the container; their concentration remains low such that the effect of inter-particle interactions is negligible. By Brownian motion a colloid arrives at a distance x (Fig. 11.5) where it experiences a force $K(x)$ due to the membrane.

Potential of mean force. Along its diffusive path, a colloid samples numerous configurations and orientations of surrounding background particles, so an ensemble of colloids experiences at x a certain mean force $K_{MF}(x)$. The corresponding potential is the *potential of mean force* $V_{MF}(x)$:

$$K_{MF}(x) = -\frac{dV_{MF}(x)}{dx}, \tag{11.15}$$

The potential $V_{MF}(x)$ is the minimal, reversible work needed to bring a colloid from infinity to position x, averaged over all configurations and orientations of the background particles. The mean force exerted on colloids in a slab with thickness dx at a distance x from the membrane with area A equals:

$$F_{MF} = K_{MF}(x)\rho(x)A dx, \tag{11.16}$$

where $\rho(x)$ is the local colloid number density, which satisfies the Boltzmann distribution:

$$\rho(x) = \rho_0 \exp[-V_{MF}(x)/kT] \tag{11.17}$$

Here ρ_0 is the bulk density at $x \to \infty$. From Eqs. (11.15) to (11.17) we find for the total force on all Brownian particles due to the membrane:

$$F_{tot} = A \int_0^\infty K_{MF}(x)\rho(x)\,dx = -A\rho_0 \int_0^\infty \frac{dV_{MF}(x)}{dx}\exp[-V_{MF}(x)/kT]\,dx \tag{11.18}$$

Boundary conditions. The two obvious limiting conditions are that any force exerted by a membrane vanishes when colloids wander off to infinity, and that colloids experience a very steep repulsion when their centers approach the membrane surface within a distance equal to the colloid radius R:

$$\frac{V_{MF}(x \to \infty)}{kT} = 0; \quad \frac{V_{MF}(x \le R)}{kT} = \infty \tag{11.19}$$

Hence (11.18) yields

$$F_{tot} = A\rho_0 kT\left(\exp[-V_{MF}(x \to \infty)/kT] - \exp[-V_{MF}(x \le R)/kT]\right) = A\rho_0 kT \tag{11.20}$$

According to Newton's third law, the total force (11.20) exerted by the membrane on the colloids, and directed into the solution, equals in magnitude the total force exercised by the colloids on the membrane. Therefore the excess pressure π exerted by Brownian particles on the plate equals:

$$\pi = \frac{F_{tot}}{A} = \rho kT, \tag{11.21}$$

which is Van 't Hoff's law.[3] It should be noted that (1) the particles in this derivation are of arbitrary size, shape and composition and (2) the force $K(x)$ exerted by the membrane remains unspecified.

What this derivation illuminates is that addition of non-interacting thermal particles increases the pressure with $\rho_0 kT$, independent of the nature of the surroundings of the particles, as long as the effect of this surroundings is averaged out in the determination of the potential of mean force. Stated differently, whenever background particles do not affect the mean force between a Brownian particle and a surface, the excess pressure is given by (11.21). Further, within this statistical approach there is no difference between the ideal gas law and Van 't Hoff's osmotic pressure law. In the former case, the background is empty space which obviously does not influence the mean force between colloid and the plate. In the latter case, the background system is a thermal molecular fluid that does not affect the mean force between dilute colloids and plate either.

Exercises

11.1 Verify that (11.14) follows from (11.11).
11.2 Where in the derivation in Sect. 11.3 is it assumed that colloids are ideal solutes?
11.3 A dispersion contains a weight concentration c_{tot} of colloids that are polydisperse in size. Show that the osmotic pressure of the dispersion yields the number averaged molecular mass M_n.
11.4 What is the weight concentration and osmotic pressure of a solution of glucose ($M = 180$ g/mol) that is isotonic with a 9 w% solution of NaCl ($M = 58$ g/mol)?

References

The derivation of Van 't Hoff's law in Section 11.3 employing a potential of mean force has not been reported earlier. The derivation was found by Prof. Agienus Vrij who handed it over to me in an unpublished document (2013) of which Section 11.3 is an adaption.

For other statistical treatments of osmotic pressure, see: G. Joos, *Lehrbuch der theoretischen Physik* (3rd ed., Blackie & Son, London, 1958); Transl. *Theoretical Physics* (Dover, 1986), pp. 594–595; P. Debye, *Théorie cinétique des lois de la pression osmotique des électrolyte forts*. Recueil des Travaux Chimiques des Pays-Bas **XLII** (1923), pp. 597–604; P. Debye in: *Topics in Chemical Physics*, Eds. A. Prock and G. Mconkey (Elsevier, Amsterdam, 1962).

An English translation of Van 't Hoff's publication on osmotic pressure in *Zeischrift für physikalische Chemie*, **vol. i.** (1887), pp. 481–508, can be found under the title *The role of osmotic pressure in the analogy between solutions and gases*, in Alembic Reprint No. 19. *The Foundations of the Theory of Dilute Solutions* (E. & S. Livingstone LTD, Edinburgh, 1961).

For a didactic treatment of the thermodynamics of dilute solutions—and thermodynamics in general—see P. Atkins, J. de Paula and J. Keeler, *Physical Chemistry* (Oxford University Press, 2014).

---

[3]The subscript '0' in ρ_0 has been dropped.

Appendix A
Moments, Fluctuations and Gaussian Integrals

Gaussians. The Maxwell-Boltzmann distributions (Fig. 3.7) from Chap. 3 and the bell-shaped solutions (Fig. 5.3) of the diffusion equation are members of a whole family of so-called Gaussian or *normal* distributions defined as

$$G(s) = \frac{1}{\sigma\sqrt{2\pi}} \exp[-\frac{1}{2}\left(\frac{\Delta s}{\sigma}\right)^2]; \quad \Delta s = s - <s> \tag{A.1}$$

Here σ is the standard deviation of the distribution:

$$\sigma = \sqrt{<s^2> - <s>^2} \tag{A.2}$$

An exponent of the form $\exp(-x^2)$ is called a Gaussian function and its integral is the Gaussian integral

$$I = \int\limits_{-\infty}^{+\infty} e^{-x^2} dx \tag{A.3}$$

This integral can be evaluated via its quadratic:

$$I^2 = \left(\int\limits_{-\infty}^{+\infty} e^{-x^2} dx\right)^2 = \int\limits_{-\infty}^{+\infty} \int\limits_{-\infty}^{+\infty} e^{-(u^2 + v^2)} du\, dv \tag{A.4}$$

This double-integral is the sum over the whole area of the u-v plane, a summation that can also be done by employing polar coordinates. On substitution of

$$r^2 = u^2 + v^2; \quad dudv = rdrd\theta, \tag{A.5}$$

© Springer Nature Switzerland AG 2018
A. P. Philipse, *Brownian Motion*, Undergraduate Lecture Notes in Physics,
https://doi.org/10.1007/978-3-319-98053-9

we obtain:

$$I^2 = \int\limits_{0}^{2\pi} \int\limits_{0}^{+\infty} e^{-r^2} r\,dr\,d\theta \overset{s=r^2}{=} 2\pi \int\limits_{0}^{+\infty} \frac{1}{2} e^{-s}\,ds = \pi \tag{A.6}$$

Hence the Gaussian integral from (A.3) equals

$$\int\limits_{-\infty}^{+\infty} e^{-x^2}\,dx = \sqrt{\pi} \tag{A.7}$$

Employing the substitution $x^2 = ay^2$, where a is a constant, the Gaussian integral modifies to:

$$\int\limits_{-\infty}^{+\infty} e^{-ay^2}\,dy = \sqrt{\frac{\pi}{a}} \tag{A.8}$$

Moments. In Chap. 3 we have calculated the first moment $<u>$ and the second moment $<u^2>$ of the distribution of particle speeds, the latter leading to the rms-speed u_{rms}. The average speed $<u>$ is numerically different from u_{rms} (see below) and we have seen that also their areas of application differ. The first moment is employed for time-dependent processes such as collision frequencies (Sect. 3.2); the second moment of the speed distribution is set into action to calculate features that involve kinetic energies, such as the pressure exerted by particles (Sect. 3.3).

Standard deviation. The width of any distribution (here of speeds u) can be quantified by its *absolute* standard deviation (SD), the quadratic of which is defined as:

$$\sigma_{\text{u}}^2 = <u^2> - <u>^2 \tag{A.9}$$

So the SD is the second moment of the speed distribution minus the square of the first moment (the average speed). The quadratic of the *relative* SD is given by:

$$\left(\frac{\sigma_{\text{u}}}{<u>}\right)^2 = \frac{<u^2>}{<u>^2} - 1 \tag{A.10}$$

The quadratic of the relative SDV must be a positive number; this necessitates the validity of the following inequality

$$\langle u^2 \rangle \geq \langle u \rangle^2 \tag{A.11}$$

In words: the average of the squared speed always exceeds the square of the average speed.

Fluctuations. It is an instructive exercise in *statistical fluctuations* to give an additional proof of the inequality in (A.11), in particular because the inequality applies to *any* statistical distribution. For the speed distribution these fluctuations are defined as the (positive or negative) differences Δu between individual particle speeds and the average speed of all particles (Fig. A.1);

$$\Delta u = u - <u> \tag{A.12}$$

This definition of fluctuations entails that their average over all particles is zero:

$$<\Delta u> = <u> - <u> = 0 \tag{A.13}$$

In terms of the fluctuations Δu in equation (A.12), the average of the squared speed can be written as:

$$<u^2> = <(<u> + \Delta u)^2>$$
$$= <u>^2 + 2<u><\Delta u> + <(\Delta u)^2> = <u>^2 + <(\Delta u)^2> \tag{A.14}$$

Since the average of the quadratic terms $(\Delta u)^2$ must be positive, it indeed follows that $<u^2> > <u>^2$; only when all speeds are the same, both averages are equal.

The moment expansion. There is a very useful approximation for computing moments from the relative SD, for narrow distributions of *arbitrary* shape.

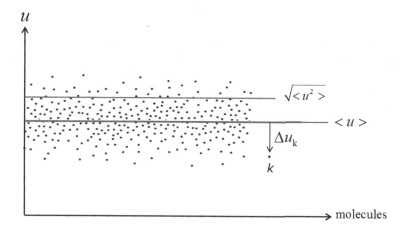

Fig. A.1 Dots represent individual particle speeds at a certain moment in time, fluctuating around the average speed $<u>$; for particle k the magnitude of the fluctuation is $\Delta u_k < 0$. Upon squaring speeds to u^2 such negative fluctuations become positive quantities such that the root-mean-square speed u_{rms} always speed

For a normalized distribution function $P(a)$ of a variable a, the nth moment of a distribution is defined by:

$$<a^n> = \int P(a)a^n da \qquad (A.15)$$

The fluctuations $\Delta = a - <a>$ satisfy :

$$\int P(a)\Delta da = 0 \qquad (A.16)$$

Dividing the nth moment by the first moment (the number average of the distribution) to the power n we obtain:

$$\frac{<a^n>}{<a>^n} = \int P(a)\left(1 + \frac{\Delta}{<a>}\right)^n da \qquad (A.17)$$

Truncating the binomial expansion after the first two terms

$$\left(1 + \frac{\Delta}{<a>}\right)^n = 1 + n\frac{\Delta}{<a>} + \frac{n(n-1)}{2}\frac{\Delta^2}{<a>^2} + \cdots, \qquad (A.18)$$

yields

$$
\begin{aligned}
\frac{<a^n>}{<a>^n} &= 1 + \frac{n(n-1)}{2}\frac{<\Delta^2>}{<a>^2} \\
&= 1 + \frac{n(n-1)}{2}\left[\frac{<a^2> - <a>^2}{<a>^2}\right]; \quad \frac{<\Delta^2>}{<a>^2} \ll 1
\end{aligned}
\qquad (A.19)
$$

So from the average value and the relative SD of a distribution one can compute all higher moments, without specifying the distribution function. Note that the linear term in (A.18) vanishes in the integration (A.17) because of (A.16). Thus for the approximation (A.19) to apply it is the *quadratic* term that must be small, as indicated in (A.19). The implication is that (A.19) is applicable to fairly broad distributions with a relative SD of, say, 10–20%.

Appendix B
Summary Vector Calculus

The following summary primarily relates to results from vector calculus that are employed in this book. A vector $\vec{v}$ can be represented by the triplet

$$\vec{v} = [v_1, v_2, v_3], \tag{B.1}$$

where v_1 is the component of the vector along axis 1. An alternative, convenient notation is:

$$\vec{v} = \sum_i \vec{\delta}_i v_i; \quad i = 1, 2, 3 \tag{B.2}$$

Here $\vec{\delta}_1, \vec{\delta}_2$ and $\vec{\delta}_3$ are unit vectors in the direction of, respectively, axes 1, 2 and 3. The scalar (or dot) product of two vectors is:

$$\vec{v}.\vec{w} = \sum_i v_i \, w_i, \tag{B.3}$$

This outcome for a dot product follows from the dot product of the unit vectors

$$\vec{\delta}_i.\vec{\delta}_j = \delta_{ij}, \tag{B.4}$$

in which δ_{ij} is the Kronecker delta; $\delta_{ij} = 0$ for $i \neq j$ and $\delta_{ij} = 1$ for $i = j$. The vector (or cross) product of $\vec{v}$ and $\vec{w}$ is also a vector, with components given by the determinant:

$$\vec{v} \times \vec{w} = \begin{vmatrix} \vec{\delta}_1 & \vec{\delta}_2 & \vec{\delta}_3 \\ v_1 & v_2 & v_3 \\ w_1 & w_2 & w_3 \end{vmatrix} \tag{B.5}$$

© Springer Nature Switzerland AG 2018
A. P. Philipse, *Brownian Motion*, Undergraduate Lecture Notes in Physics,
https://doi.org/10.1007/978-3-319-98053-9

The vector *differential operator* $\vec{\nabla}$ ('del') is defined in Cartesian coordinates as:

$$\vec{\nabla} = \sum_i \vec{\delta}_i \frac{\partial}{\partial x_i} \tag{B.6}$$

If s is a scalar function of x_1, x_2 and x_3 then its *gradient* ('grad') is:

$$\vec{\nabla} s = \sum_i \vec{\delta}_i \frac{\partial s}{\partial x_i} \tag{B.7}$$

If $\vec{v}$ is a function of the coordinates x_i then its *divergence* ('div') is the dot product:

$$\vec{\nabla}.\vec{v} = \sum_i \frac{\partial v_i}{\partial x_i} \tag{B.8}$$

The *curl* of the vector is the cross product

$$\vec{\nabla} \times \vec{v} = \begin{vmatrix} \vec{\delta}_1 & \vec{\delta}_2 & \vec{\delta}_3 \\ \frac{\partial}{\partial x_1} & \frac{\partial}{\partial x_2} & \frac{\partial}{\partial x_3} \\ v_1 & v_2 & v_3 \end{vmatrix} \tag{B.9}$$

For example, the component of curl $\vec{v}$ in the direction of $\vec{\delta}_1$ is:

$$\left[\vec{\nabla} \times \vec{v}\right]_1 = \frac{\partial v_3}{\partial x_2} - \frac{\partial v_2}{\partial x_3}$$

The *Laplacian* of a *scalar* s is the divergence of its gradient:

$$\vec{\nabla}.\vec{\nabla} s = \sum_i \frac{\partial^2}{\partial x_i^2} s = \nabla^2 s, \tag{B.10}$$

where

$$\nabla^2 = \frac{\partial^2}{\partial x_1^2} + \frac{\partial^2}{\partial x_2^2} + \frac{\partial^2}{\partial x_3^2} \tag{B.11}$$

is the Laplacian (read: 'del squared') in Cartesian coordinates. The Laplacian of a *vector field* $\vec{v}$ is defined as:

$$\nabla^2 \vec{v} = \vec{\nabla}\left(\vec{\nabla}.\vec{v}\right) - \left[\vec{\nabla} \times \left(\vec{\nabla} \times \vec{v}\right)\right] \tag{B.12}$$

This definition is valid for curvilinear as well as rectangular coordinates.

The divergence theorem. Let S be a surface (with unit outward normal $\vec{n}$), which encloses a region with volume V. Then

$$\int_S \vec{F}.\vec{n}\,dS = \int_V \vec{\nabla}.\vec{F}\,dV, \tag{B.13}$$

which is called the *divergence theorem*; its physical meaning is further explained in Chap. 5. A similar identity is:

$$\int_S p\vec{n}\,dS = \int_V \vec{\nabla}p\,dV \tag{B.14}$$

where p is a scalar function.

Spherical Coordinates (r, θ, ϕ)

For problems involving spherical symmetry such as the sphere rotation in Sect. 8.2 and the diffusion towards a spherical target in Sect. 9.3, it is convenient to work with polar coordinates r, θ, ϕ (Fig. B.1). Below is a list of vector operations in terms of these coordinates.

$$\vec{\nabla} = \vec{\delta}_r \frac{\partial}{\partial r} + \vec{\delta}_\theta \frac{1}{r}\frac{\partial}{\partial \theta} + \vec{\delta}_\phi \frac{1}{r\sin\theta}\frac{\partial}{\partial \phi} \tag{B.15}$$

$$\vec{\nabla}p = \frac{\partial p}{\partial r}\vec{\delta}_r + \frac{1}{r}\frac{\partial p}{\partial \theta}\vec{\delta}_\theta + \frac{1}{r\sin\theta}\frac{\partial p}{\partial \phi}\vec{\delta}_\phi \tag{B.16}$$

$$\vec{\nabla} \cdot \vec{u} = \frac{1}{r^2}\frac{\partial}{\partial r}\left(r^2 u_r\right) + \frac{1}{r\sin\theta}\frac{\partial}{\partial \theta}(u_\theta \sin\theta) + \frac{1}{r\sin\theta}\frac{\partial \mu_\phi}{\partial \phi} \tag{B.17}$$

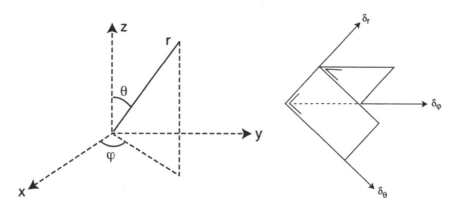

Fig. B.1 Spherical coordinates (left) and their corresponding unit vectors (right)

$$\nabla^2 p = \frac{1}{r^2}\frac{\partial}{\partial r}\left(r^2\frac{\partial p}{\partial r}\right) + \frac{1}{r^2\sin\theta}\frac{\partial}{\partial\theta}\left(\sin\theta\frac{\partial p}{\partial\theta}\right) + \frac{1}{r^2\sin^2\theta}\frac{\partial^2 p}{\partial\phi^2} \qquad \text{(B.18)}$$

$$\vec{\nabla}\times\vec{u} = \frac{1}{r^2\sin\theta}\begin{vmatrix} \vec{\delta} & r\vec{\delta}_\theta & r\sin\theta\vec{\delta}_\phi \\ \frac{\partial}{\partial r} & \frac{\partial}{\partial\theta} & \frac{\partial}{\partial\phi} \\ u_r & ru_\theta & r\sin_\theta u_\phi \end{vmatrix} \qquad \text{(B.19)}$$

$[\nabla^2\vec{u}]_\phi = \phi -$ component of the Laplacian of $\vec{u}$

$$= \frac{1}{r^2}\frac{\partial}{\partial r}\left[r^2\frac{\partial u_\phi}{\partial r}\right] + \frac{1}{r^2}\frac{\partial}{\partial\theta}\left[\frac{1}{\sin\theta}\frac{\partial u_\phi\,\sin\theta}{\partial\theta}\right] + \text{derivatives with respect to } \phi$$

$$\text{(B.20)}$$

$\tau_{r\phi} = r\phi -$ component of the stress in Newton's law

$$= -\eta\, r\frac{\partial}{\partial r}\left[\frac{u_\phi}{r}\right] + \text{derivative to } \phi \qquad \qquad \text{(B.21)}$$

Cylindrical Coordinates (r,θ,z) (Fig. B.2)

$$\vec{\nabla}p = \frac{\partial p}{\partial r}\vec{\delta}_r + \frac{1}{r}\frac{\partial p}{\partial\theta}\vec{\delta}_\theta + \frac{\partial p}{\partial z}\vec{\delta}_z \qquad \text{(B.22)}$$

$$\vec{\nabla}\cdot\vec{u} = \frac{1}{r}\frac{\partial}{\partial r}(ru_r) + \frac{1}{r}\frac{\partial u_\theta}{\partial\theta} + \frac{\partial u_z}{\partial z} \qquad \text{(B.23)}$$

$$\nabla^2 p = \frac{1}{r}\frac{\partial}{\partial r}\left(r\frac{\partial p}{\partial r}\right) + \frac{1}{r^2}\frac{\partial^2 p}{\partial\theta^2} + \frac{\partial^2 p}{\partial z^2} \qquad \text{(B.24)}$$

$[\nabla^2\vec{u}]_z = z -$ component of the Laplacian of $\vec{u}$

$$= \frac{1}{r}\frac{\partial}{\partial r}\left[r\frac{\partial u_z}{\partial r}\right] + \frac{1}{r^2}\frac{\partial^2 u_z}{\partial\theta^2} + \frac{\partial^2 u_z}{\partial z^2} \qquad \text{(B.25)}$$

Figure B.2 Cylindrical coordinates and their corresponding unit vectors

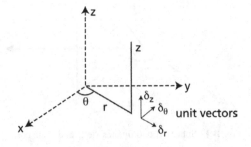

$$\tau_{rz} = rz - \text{component of the stress in Newton's viscosity law}$$

$$= -\eta \left[\frac{\partial u_r}{\partial z} + \frac{\partial u_z}{\partial r} \right] \tag{B.26}$$

References

An extensive treatment of vector (and tensor) calculus, with many applications, can be found in:
R. Bird, W. Stewart and E. Lightfoot, *Transport Phenomena* (Wiley, New York 2001), and
J. E. Marsden and A. J. Tromba, *Vector Calculus* (W. H. Freeman and Company, New York, 2003).
A useful informal text which emphasizes the physical significance of vector calculus is:
H.M. Schey, *Div, Grad, Curl, and All That*, (Norton, New York 1973)

Appendix C
Answers to Selected Exercises

Chapter 1

1.1 M: 2.5×10^{-5} cm^2/s; N: 4.9×10^{-11} m^2/s; C: 2.5×10^{-13} m^2/s; G: 2.5×10^{-16} m^2/s.

1.2. Rms displacement $= <r^2>^{1/2} = (6Dt)^{1/2}$. For an M-sphere: 0.73 cm in one hour and 68.1 cm in one year. For an N-sphere: 0.10 cm, respectively 9.7 cm.

1.3 (a) $n = 10^{15}$; (b) 10^5

1.4 For Z spheres with volume V_p in a vessel volume the sphere volume fraction is $\phi = ZV_p/V$. So the molar volume is (N_{AV} is Avogadro's number):

$$V_M = \frac{N_{AV}(4/3)\pi R^3}{0.64}$$

$M : V_M = 3.9\,\text{cm}^3/\text{mol}; N : 0.49\,\text{m}^3/\text{mol}; C : 3.94\,\text{m}^3/\text{mol};$
$G : 3.94\,\text{km}^3/\text{mol}$

Chapter 2

2.1 For a grain radius of 50 μm the displacement in one hour is: $<x^2>^{1/2} = 5.9$ μm. So you don't see grains moving when you peer through a microscope; what Brown observed was not motion of grains, but tiny particles released by them, see Sect. 2.1.

2.2 (a) See the derivation of the barometric profile in Sect. 10.5; (b) 7.9 km, respectively 50 μm, taking $T = 298$ K.

2.3 The colloid indeed dissipates motional energy to the solvent but—in equilibrium—receives on average the same kinetic energy in return.

2.4 The phenomenological Second law of thermodynamics states that no process is possible with the only result that heat withdrawn from a reservoir is fully converted to work. A bouncing ball comes to a stop when all its kinetic energy has been dissipated as heat to the surroundings; the Second Law forbids that

© Springer Nature Switzerland AG 2018
A. P. Philipse, *Brownian Motion*, Undergraduate Lecture Notes in Physics,
https://doi.org/10.1007/978-3-319-98053-9

afterwards the ball spontaneously jumps upwards by conversion of (part of) that dissipated heat to its kinetic energy. A Brownian particle, however, loses kinetic energy to its surroundings and receives on average the same quantity such that its average speed remains constant. Brownian motion, therefore, shows that the Second Law is not an absolute one: a soccer ball spontaneously jumping up from the table is not an impossible event: its occurrence is simply extremely *unlikely*. The smaller the ball, however, the more likely this event becomes: a colloidal ball randomly jumps over the table.

2.5 Lucretius does not describe Brownian motion: movements of visible dust particles in sunbeams are convective motions due to air-currents set up by temperature differences.

Chapter 3

3.1 (a) $p = 4 \times 10^{-23}$ bar.   (b) For $d \approx 10^{-10}$ m : $\lambda \approx 23 \times 10^9$ km.
(c) $<u> = 178$ m/s $\Rightarrow z = <u>/\lambda = 8 \times 10^{-12}$ s^{-1}. The time elapsed between two collisions is of order 10^6 years.

3.2 (a) $<v_f> = \frac{M_{H_2O}}{\delta \times N_{Av}} \approx 30 \times 10^{-30}$ m$^3 = 30 \left(\overset{\circ}{A} \right)^3$; $1 \overset{\circ}{A} = 10^{-10}$ m;

(b) It is not: water is hardly compressible....

(c) From the ideal gas law: $<v_f> = \frac{kT}{p} = \frac{4.12 \times 10^{-21} J (=Nm)}{10^5 Nm^{-2}} =$ $4.12 \times 10^4 \left(\overset{\circ}{A} \right)^3$.

(d) Particles in liquid water are at an average distance of about $<v_f>^{1/3} \approx 30^{1/3}$ Å $= 3.1$ Å. For the vapor: $<v_f>^{1/3} = 34.5$ Å.

3.3 $z = 1.5 \times 10^9$ s^{-1}.

3.4 $\frac{\lambda}{d} = \frac{1}{\sqrt{26}\phi} = 11.8$.

M-spheres : $\lambda = 2.4$ nm; N : 117.9 nm; C : 4.7 μm; G : 2.4 cm.

3.5 (a) $<u> = 122.5$ km/h ; (b) $<u^2>^{1/2} = 124.7$ km/h.

3.6 $<E_{kin}>$ per mole $= (3/2)kT \times N_{AV} = (3/2)RT = 3.7$ kJmol^{-1} at $T = 300$ K.

3.7 Determine the maximum of the Maxwell distribution by putting $dP(u)/du = 0$.

3.8 $<v_x^2> = \left(\frac{m}{2\pi kT} \right)^{1/2} \times \int\limits_{-\infty}^{+\infty} v_x^2 \exp[-mv_x^2/2kT]dv_x = \frac{kT}{m}$; $<v_x> = 0$.

3.10 $G(v_x) = \frac{1}{\sigma\sqrt{2\pi}} \exp[-\frac{1}{2} \left(\frac{v_x - <v_x>}{\sigma} \right)^2]$; $\sigma^2 = <v_x^2> - <v_x>^2$.

Since the distribution in v_x is symmetric, $<v_x> = 0$ so the standard deviation equals

$$\sigma^2 = <v_x^2> - <v_x>^2 = <v_x^2> = \frac{kT}{m}$$

and the Gaussian distribution becomes:

$$G(v_x) = \left(\frac{m}{2\pi kT}\right)^{1/2} \exp\left[-\frac{1}{2}\left(\frac{mv_x^2}{kT}\right)^2\right],$$

which is the normalized Maxwell-Boltzmann distribution for v_x.

Chapter 4

4.1 $<x^2>^{1/2} = 0.37\,\text{mm}$.

4.2 (a) 0.54 mm/s. (b) 0.54 mm. (c) displacement in x-direction: 2 μm.

4.4 $\tau_{CR} \sim \frac{\eta R^3}{kT}$; for M-sphere : $\tau_{CR} \sim 2 \times 10^{-13}$ s; $N : 5 \times 10^{-7}$ s; $C : 2 \times 10^{-4}$ s; $G : 2 \times 10^4$ s.

NB: these are estimates of orders of magnitude; any constant in the expression for the relaxation time is ignored here.

4.5 Each step takes τ_{MR} seconds so the number of steps is $\frac{\tau_{CR}}{\tau_{MR}} \approx \frac{2\times10^{-4}\,\text{s}}{5\times10^{-9}\,\text{s}} \approx 4 \times 10^4$, corresponding to a distance $4 \times 10^4 \ell \approx 4\,\text{μm}$; 40 times the colloid radius.

4.6
$$w = \int_0^{\tau_{MR}} fv(t)dr = f\int_0^{\tau_{MR}} v^2(t)dt = fv_0^2\int_0^{\tau_{MR}} \exp[-2t/\tau_{MR}]/dt$$

$$\approx fv_0^2\frac{\tau_{MR}}{2} = \frac{1}{2}mv_0^2$$

Comparing this work to the thermal energy we find $v_0^2 \sim 2kT/m$, which is pretty close to the equipartition result $<v^2> = 3kT/m$. So the initial kinetic energy kick $mv_0^2/2$ received by the sphere is returned as dissipated work during the momentum relaxation event.

Chapter 5

5.2 The probability to find a particle in the region from $-\sqrt{3}$ to $+\sqrt{3}$ is :

$$p = \frac{1}{\sqrt{4\pi Dt}}\int_{-\sqrt{3}}^{+\sqrt{3}} \exp[-x^2/4Dt]dx = 0.683,$$

where $Dt = (1/2) \times 3\,\text{cm}^2$, and x a distance in centimeter.

5.3 If mass is conserved then the initial particle concentration Γ_0 on the plate at $t = 0$ should equal the integrated concentration profile at any later time $t > 0$. This is indeed the case:

$$\Gamma_0 = \int_{-\infty}^{+\infty} \rho(x, t)dx = \frac{\Gamma_0}{\sqrt{4\pi Dt}} \int_{-\infty}^{+\infty} \exp[-x^2/4Dt]dx$$

$$= \frac{\Gamma_0}{\sqrt{\pi}} \int_{-\infty}^{+\infty} \exp[-y^2]dy = \frac{\Gamma_0}{\sqrt{\pi}} \sqrt{\pi} = \Gamma_0.$$

Chapter 6

6.1 $$<x^2> = \int_{-\infty}^{+\infty} P(x, t)x^2 dx = \frac{1}{\sqrt{4\pi Dt}} \int_{-\infty}^{+\infty} \exp[-x^2/4Dt]x^2 dx$$

$$= \frac{4Dt}{\sqrt{\pi}} \int_{-\infty}^{+\infty} \exp[-y^2]y^2 dy = 2Dt.$$

6.2 $\frac{d}{dt}<x^2> = \int_{-\infty}^{+\infty} \frac{d^2P}{dx^2}x^2 dx; \ P = P(x, t).$

The normalized distribution function P is a continuous, differentiable function symmetric around $x = 0$. P decays fast enough such that:

$$\lim_{x \to \pm\infty} (xP) = \lim_{x \to \pm\infty} (x^2 \frac{dP}{dx}) = 0.$$

Then:

$$\frac{d}{dt}<x^2> = D \int_{-\infty}^{+\infty} x^2 d\left(\frac{dP}{dx}\right) = -2D \int_{-\infty}^{+\infty} \frac{dP}{dx}dx^2 = -2D(-1) \int_{-\infty}^{+\infty} Pdx$$

$$= 2D \Rightarrow <x^2> = 2Dt.$$

6.3 Van 't Hoff's osmotic pressure law that underlies the force balance equation (6.17) only holds for ideal particles. In the flux balance (6.23) it is assumed that the diffusion coefficient D is independent of concentration ρ, which also presupposes non-interacting particles.

6.4 Diffusion coefficients of oxygen in water and air ($p = 1$ bar, 20 °C) are, respectively, 2.1×10^{-9} and $2.0 \times 10^{-6} \ \mathrm{m^2/s}$.

6.5 No assumption has been made on the shape of the rotating colloid; shape only enters via the choice of the rotational friction factor f_r in $D_r = kT/f_r$.

6.6 $< \cos \theta > = \int_0^\pi P(\theta, t) \sin\theta \cos\theta d\theta$. For randomly orientations $P(\theta, t) = 1$ by definition and $< \cos \theta > = -2\pi \int_0^\pi \cos\theta d(\cos\theta) = 0$

Chapter 7

7.2 Taking $L = R$, and for U the Stokes sedimentation velocity $u_{sed} = 2R^2(\delta_p - \delta)g/9\eta$ we have for the Reynolds number:

$$\text{Re} = \frac{\delta U L}{\eta} u_{sed} = \frac{2(\delta_p - \delta)\delta g}{9\eta^2} R^3 = (2.75 \times 10^{12} \, \text{m}^{-3})R^3$$

Here δ_p and δ are the mass density of, respectively, particles and water; $\eta = 0.89$ cP.

For the Particle Quartet (Table 1.1) $M : \text{Re} = 3 \times 10^{-18}$; $N : 3 \times 10^{-13}$; $C : 3 \times 10^{-9}$; $G : 3 \times 10^3$.

7.3 (a) $Re = 600.000$. (b) By far water displacement; viscous drag is unimportant because you swim at high Re. Your energy input is proportional to U^2 and not to U as in the case of viscous flow.

7.4 Re will decrease by a factor $\frac{D}{d} = \left(\frac{2D}{3L}\right)^{1/3} = 0.28$, for $\frac{L}{D} = 30$, assuming that D^2 is the frontal area facing the flow. One could argue that for flow perpendicular to the rod, Re actually increases with a factor L/D.

7.5 (a) $\frac{\partial^2 u(y)}{\partial u^2} = 0 \Rightarrow u(y) = u(D)\frac{y}{D}$; $<u> = \frac{1}{D}\int_0^D u(y)dy = \frac{1}{2}u(D)$.

(b) Stream function because $(\vec{u}.\vec{\nabla})[u(y)] = 0$.

(c) $\frac{F}{\text{area}} = -\eta\frac{\partial u(y)}{\partial y} = 10^{-3}\text{Pa}$.

Chapter 8

8.2 $\tau_{rz} = -\eta\frac{du_z}{dr} = -\frac{1}{2}\frac{dp}{dz}r = \frac{1}{2}\frac{\Delta P}{L}r$; $\Delta P = -\int_0^L \frac{dp}{dz}dz = -\frac{dp}{dz}L$.

The stress is zero at $r = 0$ and increases linearly with r to its maximum value at $r = R$. The total viscous force exerted on the tube wall is:

$$F_{vis} = 2\pi RL \times \tau_{rz}|_{r=R} = \pi R^2 \Delta P$$

Note that the flow in the tube is driven by a net external force $F_{ext} = \pi R^2 \Delta P$, exerted on the in- and outlet of the tube, which for a steady flow is exactly compensated by the total viscous force on the tube's inner surface.

8.3 $f_r = 8\pi\eta R$.

8.6 G-sphere : $U_0 = 4.9\,\text{m/s}$; C: 46 nm/s; N: 0.1 nm/s.

8.7 Volume flow rate scales as $Q \propto R^4 \Delta P \Rightarrow \frac{\Delta P_2}{\Delta P_1} = \left(\frac{R_2=5}{R_1=4}\right)^4$; increase is 244%.

Chapter 9

9.1 Take discs of equal radii R. The flux of discs in the plane towards the origin is:

$$J = 2\pi r D \frac{d\rho}{dr}$$

Here ρ is the disc number density at distance r. Since J is constant (stationary diffusion) we find:

$$J \int\limits_{\delta}^{2R} \frac{dr}{r} = 2\pi D(0 - \rho_{bulk}) \Rightarrow J = \frac{2D\pi\rho_{bulk}}{\ln(\delta/2R)}$$

One cannot, in contrast to the 3-dimensional case, take the limit $\delta \to \infty$ because then the diffusion flux would vanish. Thus for diffusion-controlled processes on a surface one has to specify a certain diffusion 'zone thickness' δ, a choice that is not required for the 3-dimensional case.

9.2 Suppose $\vec{r}_i, \vec{r}_j$ are the position vectors of two particles that start in the origin. Then the mean-square-displacement of particles relative to each other is:

$$< (\vec{r}_i - \vec{r}_j)^2 > = 6D_{ij}t$$

For independent diffusers the position vectors are uncorrelated such that the average of their dot-product is zero. Hence:

$$< (\vec{r}_i.\vec{r}_j)^2 > -2<(\vec{r}_i \cdot \vec{r}_j) > + <(\vec{r}_i \cdot \vec{r}_j)^2 > = 6D_i t + 6D_j t,$$

from which it follows that: $D_{ij} = D_i + D_j$.

9.3
$$J(j \to i) = \text{const.} \times (2 + x + x^{-1}); \quad x = \frac{R_i}{R_j}$$

$$\Rightarrow \frac{d}{dx} J(j \to i) = 0, \quad \text{for } x = 1 (\text{is a minimum}).$$

9.4 3.8×10^{-5}, 3.8×10^3 s.

9.5 In Sect. 9.3 we found that the radius R_i of sphere i (with radius R_0 at time t_0) grows by the diffusive uptake of small molecules j as:

$$R_i^2 - R_i^2 = 2D_j\phi_j(t - t_0); \quad \phi_j \approx c_{j,\infty}v_j$$

Hence:

$$\frac{dR}{dt} = \frac{A}{R} \tag{1}$$

Here A is a constant; the subscript i has been dropped. Another sphere has a larger radius $R(1 + \varepsilon)$, with $\varepsilon > 0$. For this larger sphere:

$$\frac{dR(1 + \varepsilon)}{dt} = \frac{A}{R(1 + \varepsilon)} \tag{2}$$

The change in time of ε follows from the combination of (1) and (2):

$$\frac{d\varepsilon}{dt} = \frac{A}{R^2}\left[(1 + \varepsilon)^{-1} - (1 + \varepsilon)\right]; \quad \varepsilon > 0$$

Since $(1 + \varepsilon) > (1 + \varepsilon)^{-1}$ for $\varepsilon > 0$ it follows that $d\varepsilon/dt < 0$. So for any arbitrary pair of spheres the relative size difference decreases in time; consequently the whole particle size distribution sharpens by diffusional growth.

9.6 The radius follows from

$$a_B \approx R^{-1/3} \times 71 \times 10^{-6} \times m^{4/3}$$

Here the colloid radius R has the unit of meter; the numerical factor is valid for water at $T = 298$ K with a mass density of $\delta = 1$ g/mL.

9.8 Number density decreases as

$$\frac{dc}{dt} = k_{11}c^2; \quad \frac{c}{c_0} = \frac{1}{1 + k_{11}c_0 t}$$

From the experimental data: $k_{11} \approx 5 \times 10^{-12}$ cm^3/s. For rapid flocculation the prediction is a higher rate constant: $k_{11} = 8kT/3\eta \approx 12 \times 10^{-12}$ cm^3/s. Possibly the van der Waals attraction between the silica particles at contact is not large enough to induce irreversible aggregation for every encounter between the particles.

9.9 17×10^5 s^{-1}.

Chapter 10.

10.7 (a) $P(h) = C \exp[-mgh/kT]$.
   (b) Probability to find a particle in the interval h, $h+dh$;

(c) $C = mg/kT$.

(d) $<h> = C \int\limits_0^\infty h e^{-Ch} dh = -\int\limits_0^\infty h de^{-Ch} = -\int\limits_0^\infty d(h e^{-Ch}) + \int\limits_0^\infty e^{-Ch} dh = \frac{1}{C} = \frac{kT}{mg}$

Alternatively: (d) $<h> = -C \frac{d}{dC} \int\limits_0^\infty e^{-Ch} dh = -C \frac{d}{dC} \left(\frac{1}{C}\right) = \frac{1}{C}$

(e) 9.0 km and 0.8 μm. Obviously the heavy droplets stay very much closer to the surface of the earth.

(f)
$$<h^2> = C \int\limits_0^\infty h^2 e^{-Ch} dh = -\int\limits_0^\infty h^2 de^{-Ch}$$

$$= -\int\limits_0^\infty d(h^2 e^{-Ch}) + \int\limits_0^\infty e^{-Ch} dh^2 = 0 + 2 \int\limits_0^\infty e^{-Ch} h dh$$

From (d) we can infer that $\int\limits_0^\infty e^{-Ch} h dh = \frac{1}{C^2}$

$$\Rightarrow <h^2> = \frac{2}{C^2} \Rightarrow h_{rms} = <h^2>^{1/2} = \frac{\sqrt{2}}{C} = \sqrt{2}\frac{kT}{mg}$$

So the rms-height is a factor $\sqrt{2}$ larger than the average height from (d).

Chapter 11

11.3 Van 't Hoff's law for species i with weight concentration c_i and molar mass M_i is:

$$\frac{\pi_i}{RT} = \frac{c_i}{M_i}$$

We measure: $\frac{\pi_{tot}}{RT} = \frac{c_{tot}}{<M>} = \frac{\sum c_i}{\sum M_i}$

Thus: $<M> = \frac{\sum c_i}{\sum c_i/M_i} = \frac{\sum n_i M_i}{\sum n_i} = M_n$

11.4 The osmotic pressure of the NaCl solution is

$$\pi = \frac{cRT}{M} = \frac{9\,\text{g/L} \times 2 \times 8.31\,\text{J/mol K} \times 300\,\text{K}}{58\,\text{g/mol}} = 7.74\,\text{bar}$$

So the isotonic solution has a sugar concentration of $\frac{180}{58} \times 9\,\text{g/L} = 27.9\,\text{g/L}$

Index

A

Absorber, 123
Aerosols, 21, 93
Angular distribution function, 140
Angular equilibrium profile, 90
Angular flux, 89
Angular work, 90
Angular velocity, 52, 88, 89, 109, 110
Angular momentum relaxation time, 52
Avogadro's number, 14, 17, 18, 20, 165
Avogadro's principle, 24
Axial flow, 106, 107
Azimuthal angle, 113

B

Ballistic motion, 47–49, 55, 56, 73, 84
Barometric profile, 16, 143, 144, 145, 165
Binary collisions, 27
Blood arteries, 123
Blood cells, 91, 120
Blood pressure, 108, 131
Boltzmann constant, 3, 14, 15, 17, 145
Boltzmann distribution, 14
Boltzmann factor, 34, 35, 38
Brownian collision time, 58, 129
Brownian displacements, 71, 82, 133
Brownian magnets, 7
Brownian motion, 1–7, 9, 10, 12, 17–19, 34, 42, 43, 47, 49, 53–56, 59, 61, 64, 66–68, 71, 72, 77–79, 82, 85, 88, 91, 93, 102, 105, 121–126, 129, 131, 133, 134, 136, 138–140, 142–144, 151, 166
Buys Ballot's objection, 47, 48

C

Cavity, 112, 120
Chemical atoms, 14
Clarkia Pulchella, 4, 9, 11, 19
Collective diffusion, 65
Couette geometry, 102, 105
Collision diameter, 25
Collision frequency, 26, 44, 69, 125–129, 138
Colloidal stability, 136
Colloids, 1–12, 15–19, 21, 22, 26, 28, 35, 42, 43, 47, 49, 50, 52–68, 71, 75, 77–85, 91, 93, 102, 105, 109, 122, 123, 128, 129, 131, 133, 136, 138, 140, 141–145, 147, 151–154, 165, 167, 168, 171
Configuration space, 57
Configurational relaxation time, 57, 59, 60, 121, 129, 136
Conservation law, 61, 63, 65
Constitutive equation, 61, 64, 65
Continuity equation, 61, 63, 79, 96, 103, 123
Continuous fluid, 18, 49, 93, 120
Continuous distributions, 33–35
Continuous fluid, 18, 49, 93, 120
Convection, 10, 12, 59, 66, 67, 79, 89, 121–123
Convection-diffusion equation, 67
Couette geometry, 102, 105
Coulomb repulsions, 137, 138
Covalent bonds, 43
Creeping flow, 102, 113, 114, 116
Curl, 114, 160
Cylindrical coordinates, 107, 162

D

Dalton's law, 29

Darcy's law, 64, 105, 108, 109
Delay factor, 135–138, 145
Differential operator, 95, 114, 160
Diffusional growth, 6, 123, 127, 131, 171
Diffusion coefficient, 3–5, 8, 18, 48, 56, 58, 59,
    65–70, 74–79, 81, 84, 85, 87, 88, 90, 91,
    93, 98, 112, 118–120, 124, 125, 127,
    128, 131, 135, 137, 147, 168
Diffusion equation, 6, 61, 63, 65, 80, 81, 85,
    90, 124, 155
Diffusion in dilute gas, 61, 68
Diffusive time scale, 54, 59, 77, 81, 118
Dilute solutions, 31, 71, 148, 151, 152
Dilution-work, 148, 150
Divergence theorem, 63, 96, 161
Discrete probability distribution, 31
Distribution function, 23, 31, 33–35, 37, 41,
    44, 81, 87, 140, 145, 158, 168

E
Early-phase flocculation, 128
Effusion, 40, 41
Elastic collisions, 22
Electric field, 66, 143, 144
Ensemble average, 80, 81, 83
Entropy, 13, 14, 23, 64
Equilibrium, 2, 14–19, 22, 23, 28–30, 34, 41,
    50, 52, 64, 71, 77–79, 88–90, 134–136,
    140, 142–144, 147–151, 165
Equilibrium constant, 149
Equipartition, 15, 16, 31, 41, 47, 52, 60, 83,
    167
Equipartition theorem, 41
Euler equation, 97, 99
External fields, 4, 6, 90, 121, 133, 139–141,
    143, 144
External force, 28, 50, 52, 59, 66, 67, 71, 72,
    78–80, 88, 93, 102, 109, 133, 134, 169
External torque, 88

F
Fast flocculation, 128, 138
Fick's first law, 61, 65, 67, 68, 75, 124, 125
Fick's second law, 65, 82
First law of thermodynamics, 13
Flocculation half-time, 129
Fluctuations, 5, 12, 43, 82, 127, 155, 157, 158
Fluid velocity fields, 94, 114
Flux density, 61
Force density, 78
Free volume, 23–25, 27, 44
Friction, 2–4, 6, 18, 19, 49, 50, 52, 54, 59, 60,
    67, 70, 74, 75, 77, 80, 90, 93, 103, 105,
    108, 109, 111–113, 117–120, 168

Friction factor, 4, 6, 18, 49, 50, 52, 54, 59, 67,
    74, 75, 77, 80, 90, 93, 105, 108, 109,
    111–113, 117–120, 168

G
Gas constant, 14
Gas viscosity, 13, 69, 70
Gaussian integral, 35, 36, 155, 156
Charged colloids, 142, 143
Graham's law, 40, 41
Granular particles, 4
Gravitational length, 17

H
Hagen-Poiseuille law, 108
Hydrodynamic decay time, 54
Hydrodynamic mobility, 93
Hydrogen bonds, 43

I
Ideal gas law, 26, 30, 44, 154, 166
Ideal particles, 24, 27–29, 147, 149, 168
Incompressible fluids, 63, 96, 97, 100, 148
Ion diffusion, 77, 118
Irreversible distribution, 2

K
Kinetic energy, 2, 15, 16, 18, 22, 27, 30, 35,
    36, 42, 44, 47, 49, 50, 52, 136, 165–167
Kinematic viscosity, 69, 98
Kronecker delta, 159

L
Langevin equation, 82–84
Langevin function, 141, 145
Late-stage flocculation, 129
Latex spheres, 10, 17, 18
Laplace operator, 87
Light quanta, 18
Lotus sphere, 118–120
Liquid permeability, 64, 108, 109

M
Mass memory loss, 84
Material derivative, 94, 100
Maxwell-Boltzmann distribution, 23, 38, 39,
    44, 64, 155, 167
Mean free path, 13, 20, 23, 24, 25, 26, 44, 48,
    68, 69, 70
Mean square angular displacement, 86
Mean square displacement, 60, 77, 80–84, 121,
    170
Mechanical theory of heat, 12
Molecular collision time, 49, 50, 53

Molecular reality, 13
Molecular size, 12, 13, 20
Molecular speeds, 6, 36, 40, 47
Moment expansion, 157
Moment of inertia, 52, 53
Moments of a distribution, 34
Momentum, 27, 28, 29, 49, 50, 51, 52, 53, 54, 55, 57, 58, 59, 60, 64, 69, 70, 78, 81, 82, 83, 84, 85, 97, 98, 99, 147, 167
Momentum diffusion, 69, 98
Momentum flux, 64, 78
Momentum relaxation time, 49, 50–55, 59, 81–84
Monodisperse colloids, 17

N
Navier-Stokes equation, 96, 100
Newtonian mechanics, 6, 82
Newton's second law, 27, 28, 48, 49, 52, 83
Newton's viscosity law, 99, 107, 163
Normalized probability, 32, 34, 35

O
Oil films, 13
On the Nature of the Universe, 19
Orientational decay, 86
Orientational distribution function, 87
Osmosis, 7, 105, 148, 150
Osmotic equation of state, 142
Osmotic equilibrium, 147, 149, 150
Osmotic pressure, 7, 31, 71–74, 78, 142, 147, 149–152, 154, 168, 173

P
Partial pressures, 28, 29
Particle flux, 62, 65, 66, 74–76, 78, 79, 137, 142
Particle orientation, 4, 88
Particle position, 4, 61, 72, 80, 88, 90
Péclet number, 59, 123
Plane parallel flow, 97, 105, 114
Poiseuille flow, 105, 113, 120, 122
Polar angle, 113, 140, 141
Pollen grains, 4, 9, 19
Poly-dispersity, 126
Potential energy, 7, 16, 17, 22, 34, 101–103, 133, 145, 148, 169
Potential of mean force, 153, 154
Pressure-work, 148
Probability density, 34, 35, 38, 40, 65, 80, 81, 87

R
Radial delay factor, 136

Random motions, 21, 22, 29, 30, 36, 37, 64, 81
Random walk, 133
Raoult's law, 149, 150
Rate constants, 128–132, 139, 171, 177
Reversibility, 102, 116
Reversible flow, 102
Reversible work, 148, 153
Reynolds number, 100–103, 117, 120, 169
Root-mean-square speed, 16, 157
Rotating Stokes flow, 109
Rotational angular motion, 52
Rotational diffusion coefficient, 4, 58, 87, 88, 90, 91, 112, 119, 120
Rotational relaxation time, 58, 88

S
Saturation magnetization, 53, 140–142
Scale, 17, 43, 49, 51, 54, 55, 57, 59, 77, 81–83, 85, 93, 98, 100–102, 112, 118, 121, 122, 138, 170
Second law of thermodynamics, 165
Sedimentation-diffusion equilibrium, 16, 78
Sedimentation time scales, 54, 59
Self-diffusion, 65
Semi-permeable membranes, 151, 152
Shear flow, 88, 93, 138
Shear force, 93, 97, 98, 138
Shear stress, 98, 99
Shear viscosity, 98
Silica spheres, 5, 7, 8, 131
Single-lens microscope, 9
Slip boundary, 108, 117–120
Slip parameter, 118
Smoke, 21
Soft matter, 21, 42, 43
Spherical coordinates, 87, 109, 110, 113, 161
Sphinx, 1, 9, 10, 12, 20
Standard deviation, 155, 156, 166
Statistical thermodynamics, 12, 14, 15, 17
Stationary diffusion, 67, 68, 123, 124, 170
Stationary states, 64, 79
Steady flow, 94, 100, 169
Steady state, 64, 67, 68, 124, 125, 134–137, 145
Stick boundary, 106, 111, 112, 116, 118, 119
Stokes-Einstein diffusion coefficient, 5, 18, 67, 77, 118
Stokes equation, 6, 96, 100, 103, 105–107, 110, 113, 116
Stokes flow, 100, 102, 106, 109
Stream function, 95, 103, 113–116, 169
Streamline, 94, 95, 96, 101, 113, 114, 116
Substantial derivative, 95
Susceptibility, 141, 142

T
Thermal energy, 3, 5–7, 18, 21, 43, 77, 90, 138, 141, 147, 167
Thermal motion, 3, 9, 11, 15, 18, 21, 27, 42, 93, 136, 147, 151
Thermal stability, 42
Thermodynamic cycle, 148
Transient force, 82
Translational diffusion coefficient, 3, 56, 77
Torque, 52, 54, 88, 89, 90, 109, 111, 112, 119, 140

U
Universal Maxwell-Boltzmann distribution, 39

V
Van der Waals forces, 1, 43, 136, 137, 171
Vapor pressure, 149, 150
Vector calculus, 61, 159
Velocity distributions, 31, 35
Van' t Hoff's osmotic pressure law, 31, 147, 154, 168
Viscous friction, 19, 52, 103
Viruses, 122
Volume fraction, 8, 13, 20, 22, 26, 27, 44, 47, 58, 59, 126, 127, 129, 132, 165
Vrij's approach, 151

Printed in the United States
By Bookmasters